Noninferiority Testing in Clinical Trials

Issues and Challenges

Chapman & Hall/CRC Biostatistics Series

Chapman & Hall/CRC Biostatistics Series

Noninferiority Testing in Clinical Trials

Issues and Challenges

Tie-Hua Ng

CRC Press
Taylor & Francis Group
Boca Raton London New York

CRC Press is an imprint of the
Taylor & Francis Group, an **informa** business

A CHAPMAN & HALL BOOK

CRC Press
Taylor & Francis Group
6000 Broken Sound Parkway NW, Suite 300
Boca Raton, FL 33487-2742

ISBN 13: 978-1-4665-6149-6 (hbk)

Library of Congress Cataloging-in-Publication Data

Ng, Tie-Hua, author.
  Noninferiority testing in clinical trials : issues and challenges / Tie-Hua Ng.
  p. ; cm. -- (Chapman & Hall/CRC biostatistics series)
  Includes bibliographical references and index.
  ISBN 978-1-4665-6149-6 (hardback : alk. paper)
  I. Title. II. Series: Chapman & Hall/CRC biostatistics series (Unnumbered)
  [DNLM: 1. Controlled Clinical Trials as Topic--methods. 2. Data Interpretation, Statistical. 3. Therapeutic Equivalency. WA 950]

  R853.C55
  615.5072′4--dc23   2014036257

Visit the Taylor & Francis Web site at
http://www.taylorandfrancis.com

and the CRC Press Web site at
http://www.crcpress.com

To Moh-Jee

Contents

Preface

An experimental (or test) treatment is often compared to an active control (or standard of care) instead of a placebo control in a randomized controlled trial for ethical reasons, with the objective to show that the experimental treatment is not inferior to the active control. The number of research articles on noninferiority (NI) testing in clinical trials has increased exponentially over the past 15 years. Five special issues on NI were published by three journals in the mid-2000s: two by *Statistics in Medicine* (2003, 2006), two by *Journal of Biopharmaceutical Statistics* (2004, 2007), and one by the *Biometrical Journal* (2005). One additional special issue on NI trials was published by *Biopharmaceutical Statistics* in 2011. In addition, there were two workshops on NI trials in the 2000s: PhRMA (Pharmaceutical Research and Manufacturers of America) Non-Inferiority Workshop in 2002 and the U.S. FDA (Food and Drug Administration)/Industry Workshop in 2007. Last, but not least, numerous short courses and tutorials on NI trials were offered at professional meetings, including four given by me (see the list given in this preface below). The earlier ones were offered at the ICSA (International Chinese Statistical Association) Applied Statistics Symposium in 2002 and the FDA/Industry Workshop in 2003. These tremendous research efforts in NI trials in the past 15 years have made many advances in this area.

There were two regulatory guidelines on NI trials: one titled Guideline on the Choice of the Non-Inferiority Margin, issued by the European Agency for the Evaluation of Medicinal Products, Committee for Proprietary Medicinal Products (EMEA/CPMP) in 2005, and another titled Guidance for Industry on Non-Inferiority Clinical Trials, issued by the FDA in 2010. The latter is in draft form and has yet to be finalized.

A recent book by Rothmann, Wiens, and Chan titled *Design and Analysis of Non-Inferiority Trials* was published in 2012. It provides (1) a comprehensive discussion on the purpose and issues involved in NI trials and (2) a thorough discussion on the most important aspects involved in the design and analysis of NI trials.

This book is intended for statisticians and nonstatisticians involved in drug development, and it is written to be easily understood by a broad audience without any prior knowledge of NI testing. It focuses on the rationale of choosing the NI margin as a small fraction of the therapeutic effect of the active control in a clinical trial. The book discusses issues with estimating the effect size based on historical placebo control trials of the active control. It also covers basic concepts related to NI trials, such as assay sensitivity, constancy assumption, discounting, and preservation. Other topics described include patient populations, three-arm trials, and equivalence of three or more groups.

Chapter 1 sets the stage for the book and ends with notations of formulation of the NI hypothesis. Chapter 2 elaborates on the rationale of choosing the NI margin as a small fraction of the therapeutic effect of the active control in a clinical trial based on the mean difference with a continuous endpoint. This chapter also covers basic concepts related to NI trials, such as assay sensitivity, constancy assumption, discounting, and preservation. Chapter 3 deals with NI testing based on (1) the mean ratio with a continuous endpoint and (2) the hazard ratio with a survival endpoint. Chapter 4 deals with NI trials with a binary endpoint. One of the strengths of this book is the use of the same notations, regardless of whether the endpoint is continuous, binary, or time-to-event. The major references for the first four chapters are the two papers published in *Drug Information Journal* in 1993 and 2001 and another paper published in *Statistics in Medicine* in 2008. These three papers were written by me.

Chapter 5 discusses (1) two statistical approaches: the fixed-margin method and the synthesis method and (2) the concepts of preservation and discounting. Chapter 6 discusses controversial issues with switching between superiority and NI. The major references for this chapter are the two papers published in the *Journal of Biopharmaceutical Statistics* in 2003 and 2007. These two papers are also written by me. Chapter 7 covers issues with estimating effect size based on multiple historical placebo-controlled trials of the active control. Chapter 8 discusses (1) the gold-standard NI design that includes a placebo group and (2) equivalence of three test groups. Chapters 9, 10, and 11 discuss regulatory guidelines, the analysis data set, and the thrombolytic example, respectively. Finally, Chapter 12 highlights (1) the fundamental issues, (2) advances, (3) current controversial issues, and (4) issues and challenges, in the design and analysis of NI trials.

This book can be read fully in the order presented. However, readers may skip Chapters 6, 8, and 9 without losing the continuity of the materials. The contents of Chapters 1 through 6 have been covered by the half-day short course/tutorial presented by me at the following conferences:

- The 45th DIA Annual Meeting, San Diego, June 2009
- The 65th Annual Deming Conference on Applied Statistics, Atlantic City, December 2009
- The 20th ICSA Applied Statistics Symposium, New York, June 2011
- The 24th Annual EuroMeeting, Copenhagen, March 2012

I appreciate the assistance of the reviewers of this book and the book proposal for their careful and insightful review. I am also indebted to David Grubbs and many others at Taylor & Francis Publishing for their guidance and patience.

I thank Chunrong Chen, Mary Lin, Lei Nie, Anthony Orencia, Jordi Ocaña Rebull, and Thamban Valappil for their review and helpful comments of early drafts of the manuscript; Thamban Valappil for the discussion in Section 7.6; Marc Walton for providing the data used in the meta-analysis conducted by CBER (Center for Biologics Evaluation and Research Center) in 1998 (Section 11.3.2); Ghanshyam Gupta for the discussion on the thrombolytic example in Chapter 11; Mark Rothmann for providing electronic copies of many references; and Moh-Jee Ng for her assistance in literature searches.

I am grateful for the support and encouragement provided by my family. My deepest gratitude goes to my wife, Moh-Jee, for her patience and support in the preparation of this book.

1

Introduction

1.1 Equivalence and Noninferiority Testing

There are two major types of equivalence in clinical research: therapeutic equivalence and bioequivalence. Therapeutic equivalence is sometimes referred to as clinical equivalence, and often arises in active-control equivalence studies in which an experimental or test treatment is compared to a standard therapy or active control based on clinical endpoints. The objective is to show that the experimental treatment produces the same benefit as the active control. The term "active-control equivalence studies" (ACES) is attributed to Makuch and Johnson (1990).

Bioequivalence arises from studies in which a test product is compared to a reference product with respect to pharmacokinetic parameters, such as the area under the concentration-time curve (AUC), the maximum concentration (C_{max}), etc. The objective of the bioequivalence studies is to show that the pharmacologic activity of one product is similar to that of another. These studies are often conducted with normal, healthy volunteers using the standard 2×2 crossover design. Chow and Liu (2009, p. 1) state, "When two formulations of the same drug or two drug products are claimed to be bioequivalent, it is believed that they will provide the same therapeutic effect or that they are therapeutically equivalent and they can be used interchangeably." This assumption is the basis for the approval of generic drugs by the U.S. Food and Drug Administration (FDA) and by nearly all regulatory agencies in the world. Temple (1982) raised a fundamental issue with regard to ACES. He questioned whether the positive control in an ACES would have beaten a placebo group had one been present. This is the assay sensitivity of the trial as discussed in the International Conference on Harmonization (ICH) E10 (2001). ICH E10 (2001, p. 7) states, "Assay sensitivity is a property of a clinical trial defined as the ability to distinguish an effective treatment from a less effective or ineffective treatment." Assay sensitivity cannot be validated in an ACES with no concurrent placebo. Without assay sensitivity, the results obtained from an ACES are uninterpretable. Therefore, a concurrent placebo is often recommended if placebo use is ethical in the setting of the study. Assay sensitivity is the fundamental issue in an ACES and will be discussed further in Sections 2.6 and 2.7.

Even though a placebo does not play any role in bioequivalence studies, these studies have issues similar to assay sensitivity. These issues, however, have received much less attention than those with ACES. On the other hand, in bioequivalence studies for biological products, such as immunoglobulin, the endogenous level, for example, needs to be taken into account by subtracting it from the concentration before computing the AUC (Ng 2001; EMEA/CPMP 2010), and the endogenous level, in some sense, plays the role of placebo.

"For drugs, bioequivalence studies are the basis for evaluating generic products. For biologics, these studies are conducted to show comparability of production lots when the sponsor makes significant manufacturing changes, such as scaling up pilot plant production or building new facilities that do not require efficacy studies or extensive safety data" (Ng 2001, p. 1518). The *generic versions* of biologic products are usually referred to as (1) biosimilars by the European Medicines Agency (EMA) of the European Union (EU) and (2) follow-on biologics by the U.S. FDA. Due to the complexity of biosimilar drug products, "the design and analysis to evaluate the equivalence between the biosimilar drug product and innovator products are substantially different from those of chemical generic products" (Chow 2011, p. 6).

Equivalence testing can arise in many other situations. One such situation is in the evaluation of product quality, such as red blood cells, plasma, and platelets. Other examples include lot-to-lot consistency and bridging studies, as well as safety studies (see also Wellek 2010). There is an extensive literature on bioequivalence. See, for example, Chow and Liu (2009), Hauschke, Steinijans and Pigeot (2007), and the references therein.

This book will focus on therapeutic equivalence. Unless otherwise noted, the discussion in this book is in the context of therapeutic equivalence, although the statistical methodology, such as the two one-sided tests procedure (Schuirmann 1987), had originally been proposed for bioequivalence.

1.2 One-Sided and Two-Sided Hypotheses

Equivalence testing with two treatment groups can be one sided or two sided. One-sided equivalence studies are also known as NI studies. Therapeutic equivalence is often one sided. That is, we wish to know if the experimental treatment is not worse than the active control. Bioequivalence with biologic products can also be one sided because we are not concerned if the test product is more bioavailable than the reference product; however, that does not preclude a two-sided bioequivalence testing. On the other hand, bioequivalence with drugs is two sided, since greater bioavailability may post a safety concern (e.g., adverse events). Therapeutic equivalence can also be two sided. For example, when comparing a twice-a-day to a once-a-day regimen, a difference in either direction is worthy of note. In the remainder of this section, the hypotheses are formulated, without loss of generality, using a continuous outcome as an example.

For two-sided equivalence testing, it is not possible to show absolute equality of the two means (Ng 2001; Piaggio et al. 2006; Senn 2007, 240; Gülmezoglu et al. 2009; Suda et al. 2011) (see also Section 1.6). Therefore, hypotheses are formulated for showing δ-equivalence, a terminology introduced by Ng (1993a, 1995); that is, the absolute mean difference being less than a prespecified $\delta > 0$. More specifically, the null hypothesis

$$H_0^{(2)}: |\mu_t - \mu_s| \geq \delta$$

is tested against the alternative hypothesis

$$H_1^{(2)}: |\mu_t - \mu_s| < \delta$$

where μ_t and μ_s denote the mean response for the test (or experimental) treatment and the standard therapy (or active control), respectively. The superscript (2) indicates testing for two-sided equivalence. The two one-sided tests procedure (Schuirmann 1987; Ng 1993b; Hauschke 2001) and the confidence interval approach are often used for testing this null hypothesis (Ng 2001). Briefly, the previous hypotheses may be rewritten as

$$H_0^{(2)}: \mu_t - \mu_s \leq -\delta \text{ or } \mu_t - \mu_s \geq \delta$$

versus

$$H_1^{(2)}: \mu_t - \mu_s > -\delta \text{ and } \mu_t - \mu_s < \delta$$

which can be decomposed into two one-sided hypotheses as

$$H_{01}: \mu_t - \mu_s \leq -\delta$$

versus

$$H_{11}: \mu_t - \mu_s > -\delta$$

and

$$H_{02}: \mu_t - \mu_s \geq \delta$$

versus

$$H_{12}: \mu_t - \mu_s < \delta$$

If both H_{01} and H_{02} are rejected at a significance level of α, then $H_0^{(2)}$ will be rejected at the same significance level. This amounts to rejecting $H_0^{(2)}$ if the $100(1 - 2\alpha)\%$ [not $100(1 - \alpha)\%$] confidence interval for the mean difference completely falls inside the interval $(-\delta, \delta)$.

For average bioequivalence (bioequivalence in average bioavailability in terms of pharmacokinetic parameters such as AUC and C_{max}; Chow and Liu 2000), the hypotheses are formulated as the mean ratio with an equivalence range of 0.8 to 1.25 rather than as the mean difference. The FDA has adopted this equivalence range for a broad range of drugs based on a clinical judgment that a test product with bioavailability measures (e.g., AUC and C_{max}) outside this range should be denied market access (U.S. FDA 2001a). The FDA also recommends that the analyses of bioavailability measures be performed on a log scale; that is, the data is log transformed. Analyzing the natural log-transformed data results in testing the hypotheses based on the mean difference with an equivalence range from −0.223 to 0.223, as $\log_e(0.80) = -0.223$ and $\log_e(1.25) = 0.223$. Significance levels of 0.05 and 0.025 are typically used for the average bioequivalence and therapeutic equivalence, respectively.

For one-sided equivalence testing, assuming a larger response corresponds to a better outcome, $\mu_t - \mu_s \geq 0$ means that the test treatment is at least as good as the standard therapy, or equivalently, the test treatment is not inferior to the standard therapy. However, it is not possible to show NI literally in the sense that $\mu_t - \mu_s \geq 0$, as shown in the following.

To show NI literally, we test the null hypothesis

$$H_0: \mu_t - \mu_s < 0 \tag{1.1}$$

against the alternative hypothesis

$$H_1: \mu_t - \mu_s \geq 0$$

so that NI can be concluded when the null hypothesis is rejected. To show superiority, we test the null hypothesis

$$H_0: \mu_t - \mu_s \leq 0 \tag{1.2}$$

against the alternative hypothesis

$$H_1: \mu_t - \mu_s > 0$$

The only difference between these two sets of hypotheses is the boundary 0. More specifically, the boundary 0 is included in the null hypothesis for testing for superiority, but not for NI. Since taking the supremum of the test statistic over the parameter space under the null hypothesis for testing NI in Equation 1.1 is equal to that over the parameter space under the null hypothesis for testing superiority in Equation 1.2, testing for NI would be the same as testing for superiority. Therefore, instead of showing NI literally (i.e., $\mu_t - \mu_s \geq 0$), hypotheses are formulated to show that the experimental treatment is δ-no-worse than the active control, a terminology introduced by Ng (1993a, 1995); that is, the experiment treatment is not worse than the active control by a prespecified δ (> 0) or more. Therefore, for NI, we test the null hypothesis

$$H_0^{(1)}: \mu_t - \mu_s \leq -\delta$$

against the alternative hypothesis

$$H_1^{(1)}: \mu_t - \mu_s > -\delta$$

where the superscript $^{(1)}$ indicates testing for the one-sided equivalence (Ng 2001).

1.3 Equivalence Margin δ

What is δ? δ is the usual term for the equivalence or NI margin. The following is a list of definitions of δ found in the literature. It is by no means a complete list. Some authors used other notations, such as Θ_0 and Δ instead of δ. This list is expanded from an original 13-item list that was prepared by the author of this book for an invited talk at the Drug Information Association workshop held in Vienna, Austria, in 2001 (Ng 2001).

1. "An equivalence margin should be specified in the protocol; this margin is the largest difference which can be judged as being clinically acceptable and ..." (ICH E9 1998).

2. "This margin is the degree of inferiority of the test treatments to the control that the trial will attempt to exclude statistically" (ICH E10 2001).

3. "Choice of a meaningful value for Θ_0 is crucial, since it defines levels of similarity sufficient to justify use of the experimental treatment" (Blackwelder 1998).

4. "...that a test treatment is not inferior to an active treatment by more than a specified, clinically irrelevant amount (Noninferiority trials)..." (Hauschke, Schall, and Luus 2000).

5. "...but is determined from the practical aspects of the problem in such a way that the treatments can be considered for all practical purposes to be equivalent if their true difference is unlikely to exceed the specified Δ" (Dunnett and Gent 1977).

6. "In a study designed to show equivalence of the therapies, the quantity δ is sufficiently small that the therapies are considered

equivalent for practical purposes if the difference is smaller than δ (Blackwelder 1982).

7. "An objective of ACES is the selection of the new treatment when it is not worse than the active control by more than some difference judged to be acceptable by the clinical investigator" (Makuch and Johnson 1990).

8. "Hence, if a new therapy and an accepted standard therapy are not more than irrelevantly different concerning a chosen outcome measure, both therapies are called therapeutically equivalent" (Windeler and Trampisch 1996).

9. "The δ is a positive number that is a measure of how much worse B could be than A and still be acceptable" (Hauck and Anderson 1999).

10. "For regulatory submissions, the goal is to pick the allowance, δ, so that there is assurance of effectiveness of the new drug when the new drug is shown to be clinically equivalent to an old drug used as an active control. For trials of conservative therapies, the δ represents the maximum effect with respect to the primary clinical outcome that one is willing to give up in return for the other benefits of the new therapy" (Hauck and Anderson 1999).

11. "…where δ represents the smallest difference of medical importance. …These approaches depend on the specification of a minimal difference δ in efficacy that one is willing to tolerate" (Simon 1999).

12. "The noninferiority/equivalence margin, δ, is the degree of acceptable inferiority between the test and active control drugs that a trial needs to predefine at the trial design stage" (Hwang and Morikawa, 1999).

13. "In general, the difference δ should represent the largest difference that a patient is willing to give up in efficacy of the standard treatment C for the secondary benefits of the experimental treatment E" (Simon 2001).

14. "A margin of clinical equivalence (Δ) is chosen by defining the largest difference that is clinically acceptable, so that a difference bigger than this would matter in practice" (EMEA/CPMP 2000).

15. "To determine whether the two treatments are equivalent, it is necessary first to identify what is the smallest difference in 30-day mortality rates that is clinically important" (Fleming 2000).

16. "The inherent issue in noninferiority and equivalence studies is the definition of what constitutes a clinically irrelevant difference in effectiveness" (Hauschke 2001).

17. "The smallest value that would represent a clinically meaningful difference, or the largest value that would represent a clinically meaningless difference" (Wiens 2002).

18. "Here, M is the non-inferiority margin, that is, how much C can exceed T with T still being considered noninferior to C (M > 0)" (D'Agostino, Massaro, and Sullivan 2003).

19. "…one sets out to arbitrarily choose this minimum clinically relevant difference, commonly called delta…" (Pocock 2003).

20. "The selection of an appropriate non-inferiority margin delta (Δ), i.e., the quantitative specification of an 'irrelevant difference' between the test and the standard treatment, poses a further difficulty in such trials" (Lange and Freitag 2005).

21. "Noninferiority margin is a response parameter threshold that defines an acceptable difference in the value of that response parameter between the experimental treatment and the positive control treatment as the selected comparator. This margin is completely dictated by the study objective" (Hung, Wang, and O'Neill 2005).

22. "If δ represents the degree of difference we wish to rule out, then we test H_{0A}: $\tau \leq -\delta$ against H_{1A}: $\tau > -\delta$ and H_{0B}: $\tau \geq \delta$ against H_{1B}: $\tau < \delta$" (Senn 2007, 238).

23. "Because proof of exact equality is impossible, a prestated margin of noninferiority (Δ) for the treatment effect in a primary patient outcome is defined" (Piaggio et al. 2006).

24. "Because proof of exact equality is impossible, a prestated margin of noninferiority (Δ) for the difference in effectiveness has to be defined" (Gülmezoglu et al. 2009).

25. "Since it is not possible to determine that the drugs being compared are exactly equal, a margin of noninferiority is determined a priori and is used to demonstrate the relative effect of the study intervention" (Suda et al. 2011).

Most definitions relate δ to a clinical judgment; others relate δ to other benefits. The ICH E10 document (2001) refers to δ as the degree of inferiority of the test treatments to the control that the trial will attempt to exclude statistically. It says exactly what the statistical inference does. The document then states that "if the confidence interval for the difference between the test and control treatments excludes a degree of inferiority of the test treatment that is as large as, or larger than, the margin, the test treatment can be declared noninferior." There is no problem with this statement if δ is small, but it could be misleading if δ is too large (Ng 2001).

In the 1990s, most ACES were not recognized as one-sided versions of equivalence. For example, ICH E10 (2001, p. 7) states the following:

> Clinical trials designed to demonstrate efficacy of a new drug by showing that it is similar in efficacy to a standard agent have been called *equivalence* trials. Most of these are actually noninferiority trials, attempting

to show that the new drug is not less effective than the control by more than a defined amount, generally called the margin.

ICH E9 (1998) distinguishes the two-sided equivalence as *equivalence* and one-sided equivalence as NI by stating that "...This type of trial is divided into two major categories according to its objective; one is an *equivalence* trial (see Glossary) and the other is a *noninferiority* trial (see Glossary)." Even so, the NI margin is not used in ICH E9 (1998), and the lower equivalence margin is used instead. This is in contrast to two more recent regulatory guidances (EMEA/CPMP, 2005; U.S. FDA, 2010), where NI is the primary focus. However, these documents do not define explicitly what δ is, as shown by the following statements made by the EMEA/CPMP (2005, p. 3) and U.S. FDA (2010, p. 7), respectively; neither do many authors (e.g., Ng 2001; Piaggio et al. 2006; Senn 2007, 238; Ng 2008; Gülmezoglu et al. 2009; Suda et al. 2011):

> In fact a noninferiority trial aims to demonstrate that the test product is not worse than the comparator by more than a pre-specified, small amount. This amount is known as the noninferiority margin, or delta (Δ).
>
> ...the NI study seeks to show that the difference in response between the active control (C) and the test drug (T), (C-T)....is less than some pre-specified, fixed noninferiority margin (M).

1.4 Choice of δ

How do you choose δ? The following is a list of suggestions in the literature. It is by no means a complete list. Again, this list is expanded from an original 9-item list that was prepared by the author of this book for an invited talk at the Drug Information Association workshop held in Vienna, Austria, in 2001 (Ng 2001).

1. "...should be smaller than differences observed in superiority trials of the active comparator" (ICH E9 1998).

2. "The margin chosen for a noninferiority trial cannot be greater than the smallest effect size that the active drug would be reliably expected to have compared with placebo in the setting of the planned trial. ...In practice, the noninferiority margin chosen usually will be smaller than that suggested by the smallest expected effect size of the active control because of interest in ensuring that some clinically acceptable effect size (or fraction of the control drug effect) was maintained" (ICH E10 2001).

3. "Θ_0 must be considered reasonable by clinicians and must be less than the corresponding value for placebo compared to standard

treatment, if that is known. ...The choice of Θ_0 depends on the seriousness of the primary clinical outcome, as well as the relative advantages of the treatments in considerations extraneous to the primary outcome" (Blackwelder 1998).

4. "In general, the equivalence limits depend upon the response of the reference drug" (Liu 2000a).

5. "On the other hand, for one-sided therapeutic equivalence, the lower limit L may be determined from previous experience about estimated relative efficacy with respect to placebo and from the maximum allowance which clinicians consider to be therapeutically acceptable. ...Therefore, the prespecified equivalence limit for therapeutic equivalence evaluated in a noninferiority trial should always be selected as a quantity smaller than the difference between the standard and placebo that a superior trial is designed to detect" (Liu 2000b).

6. "The extent of accepted difference (inferiority) may depend on the size of the difference between standard therapy and placebo" (Windeler and Trampisch 1996).

7. "A basis for choosing the δ for assurance of effectiveness is prior placebo-controlled trials of the active control in the same population to be studied in the new trial" (Hauck and Anderson 1999).

8. "This margin chosen for a noninferiority trial should be smaller (usually a fraction) than the effect size, Δ, that the active control would be reliably expected to have compared with placebo in the setting of the given trial" (Hwang and Morikawa 1999).

9. "The difference δ must be no greater than the efficacy of C relative to P and will in general be a fraction of this quantity delta δ_c." (Simon 2001).

10. "Under other circumstances it may be more acceptable to use a delta of one half or one third of the established superiority of the comparator to placebo, especially if the new agent has safety or compliance advantages" (EMEA/CPMP 1999).

11. "Choosing the noninferiority margin M ... we need to state the noninferiority margin M, that is, how close the new treatment T must be to the active control treatment C on the efficacy variable in order for the new treatment to be considered noninferior to the active control" (D'Agostino, Massaro, and Sullivan 2003).

12. "The size of the acceptable margin depends on the smallest clinically significant difference (preferably established by independent expert consensus), expected event rates, the established efficacy advantage of the control over placebo, and regulatory requirements" (Gomberg-Maitland, Frison, and Halperin 2003).

13. "A prestated margin of noninferiority is often chosen as the small-est value that would be a clinically important effect. If relevant, Δ should be smaller than the 'clinically relevant' effect chosen to investigate superiority of reference treatment against placebo" (Piaggio et al. 2006).

14. "The choice of the noninferiority margin can be made using clinical assessment, which is to a certain extent arbitrary, and needs consensus among different stakeholders. ... A reasonable criterion is to preserve 80% of the benefit of the full AMTSL package (considered as 100%) over expectant management (considered as 0%). ...Preserving a higher percentage (say 90%) will push the sample size calculations very high while a smaller percentage (say 50%) may not be considered acceptable" (Gülmezoglu et al. 2009).

15. "The margin of noninferiority is typically selected as the smallest value that would be clinically significant ..." (Suda et al. 2011).

16. "The determination of the noninferiority margin should incorporate both statistical reasoning and clinical judgment" (Suda et al. 2011).

Ng (1993b, 2001, 2008) proposed that the equivalence margin δ should be a small fraction (e.g., 0.2) of the therapeutic effect of the active control as compared to placebo (or effect size). This proposal is in line with the view of many authors that δ should depend on the effect size of the active control, but more specifically, recommends that δ should be a small fraction of the effect size with the objective of showing NI. Such a proposal and its motivation will be elaborated in Chapter 2.

ICH E10 (2001) and EMEA/CPMP (2005) suggested that the determination of δ be based on both statistical reasoning and clinical judgment, which was supported by many authors (e.g., Kaul and Diamond 2007; Suda et al. 2011). The "statistical reasoning" is due to the dependence of δ on the effect size, as Suda et al. (2011) stated that "statistical reasoning takes into account previous placebo-controlled trials to identify an estimate of the active control effect." There is a subtle difference in wording with regard to "clinical judgment." The following are two lists of such wordings:

1. Clinically acceptable difference, clinically irrelevant amount, clinically irrelevant difference, clinically meaningless difference, irrelevantly different, degree of acceptable inferiority

2. Clinically important, clinically meaningful difference, clinically relevant difference, clinically significant

The key word in the first and second list is "acceptable" and "important," respectively. There are two approaches to set δ: (1) bottom-up and (2) top-down. The bottom-up approach starts from the bottom with an extremely small difference that is "acceptable" and works up, while the top-down approach starts from the top with a large difference that is "important" and works down. Hopefully, these two approaches stop at the same place, which becomes the δ. Treadwell et al. (2012, p. B-5) stated the following:

> The two journal publications (Gomberg-Maitland, Frison, and Halperin 2003; Piaggio et al. 2006) described the threshold in terms of the *smallest value that would be clinically important.* Three regulatory documents (ICH E9 1998; EMEA/CPMP 2000; US FDA 2010) described it as the *largest difference that is clinically acceptable.*

The first sentence corresponds to the top-down approach, while the second sentence corresponds to the bottom-up approach. See Section 2.8 of Chapter 2 for further discussion of these approaches when δ is expressed in terms of a fraction of effect size of the standard therapy as compared to placebo.

1.5 Ethical Issues and Declaration of Helsinki

1.5.1 Gold Standard for Assessing Treatment Efficacy

A randomized, double-blind, placebo-controlled trial is the gold standard in assessing the efficacy of the test treatment. In such a trial, subjects are randomly assigned to either the test treatment or the placebo. "The purpose of randomization is to avoid selection bias and to generate groups which are comparable to each other" (Newell 1992, p. 837). Without randomization, the investigators can preferentially (intentionally or unintentionally) enroll subjects between the two groups. In addition, unobservable covariates that affect the outcome are most likely to be equally distributed between the two groups; thus, it minimizes allocation bias. Double-blind means both the investigator and the participant are unaware of the treatment (test treatment or placebo) the participant is receiving. Without double-blinding, the results may be subjected to potential bias, especially if the outcome variable is subjective. It is critical that randomization be properly executed, and blinding is adequate because lack of proper randomization and/or inadequate blinding may render the results uninterpretable due to various biases, which are difficult, if not impossible, to assess and account for. With a proper randomization and adequate blinding, any observed difference beyond the random chance may then be attributed to the test treatment.

The analysis of a placebo-controlled trial is relatively simple and straightforward as compared to an active-controlled trial. This will be elaborated in Section 1.5.3. However, when effective treatment is available, placebo-controlled trials are under attack for ethical reasons (Lasagna 1979). The Declaration of Helsinki calls for using the best currently proven intervention as the control (see Section 1.5.2). This makes a lot of sense from an ethical point of view. However, such a recommendation was pushed back notably by the regulatory agency (see Section 1.5.2) due to inherent difficulties in the interpretation of an ACES (see Section 1.5.3; Temple 1997).

1.5.2 Declaration of Helsinki and U.S. Regulations

A brief overview of the Declaration of Helsinki is given by Wikipedia Contributors (2014) in the following:

> The Declaration of Helsinki is a set of ethical principles regarding human experimentation developed for the medical community by the World Medical Association (WMA). The declaration was originally adopted in June 1964 in Helsinki, Finland, and has since undergone seven revisions (the most recent at the General Assembly in October 2013) and two clarifications, growing considerably in length from 11 paragraphs in 1964 to 37 in the 2013 version.

Article II.3 in the 3rd Revision of the Declaration (WMA 1989) stated:

> In any medical study, every patient—including those of a control group, if any—should be assured of the best proven diagnostic and therapeutic method.

The following was added to the end of that statement in the 4th Revision of the Declaration (WMA 1996):

> This does not exclude the use of inert placebo in studies where no proven diagnostic or therapeutic method exists.

Article 32 in the 6th Revision (WMA 2008) of the Declaration stated:

> The benefits, risks, burdens and effectiveness of a new intervention must be tested against those of the best current proven intervention, except in the following circumstances:
>
> - The use of placebo, or no treatment, is acceptable in studies where no current proven intervention exists; or
> - Where for compelling and scientifically sound methodological reasons the use of placebo is necessary to determine the efficacy or safety of an intervention and the patients who receive placebo or no treatment will not be subject to any risk of serious or irreversible harm. Extreme care must be taken to avoid abuse of this option.

A minor revision of this article was made in the current 7th Revision (WMA 2013) of the Declaration as Article 33 under "Use of Placebo."

In 1975, the U.S. FDA incorporated the 1964 Helsinki Declaration into its regulation governing investigational drug trials conducted in non-U.S.

countries (U.S. FDA 2001b). The agency also issued a similar regulation applicable to devices in 1986, when the 1983 version of the declaration (2nd Revision) was in effect (21 CFR 814.15). Subsequently, the agency amended the regulation in 1981 to replace the 1964 declaration with the 1975 version (1st Revision), and again in 1991 (21 CFR 312.120) to replace the 1975 declaration with the 1989 version (3rd Revision). The regulations (21 CFR 312.120 and 21 CFR 814.15) have not been amended to incorporate the 2000 version of the declaration (5th Revision) (U.S. FDA 2001b), and it is silent with regard to the 1996 version of the declaration (4th Revision). On April 28, 2008, the regulations were amended, again abandoning the Declaration of Helsinki. Instead, it is required to follow the ICH E6 (1996) guidance on good clinical practice (GCP) such as review and approval by an independent ethics committee (IEC) and informed consent from subjects. This requirement took effect on October 27, 2008, and was codified in 21 CFR 312.120 (U.S. FDA 2012).

1.5.3 Placebo Control versus Active Control and Sample Size Determination

For a placebo-controlled trial, the null hypothesis is that there is no difference between test treatment and placebo. This null hypothesis is usually tested at a two-sided 0.05 or one-sided 0.025 significance level. Such hypothesis testing has been referred to as the *conventional approach* (Ng 1995). See Section 1.6 for further discussion of using this approach in an ACES. Aside from randomization and blinding, there is an incentive to conduct high-quality studies, as poorly conducted studies (e.g., mixing up treatment assignment, poor compliance, etc.) may not detect a true difference, since there is an increase in variability and bias toward equality. Despite the many scientific merits of placebo-controlled trials, such trials have been controversial from an ethical standpoint as increasing number of effective treatments are brought into the market. For example, the Declaration of Helsinki (1989) (see Section 1.5.2) essentially called for an active-controlled trial rather than a placebo-controlled trial when effective treatments are available.

For an active-controlled trial, if the objective is to show superiority of test treatment over the active control, the statistical principle in hypothesis testing is the same as that in the placebo-controlled trial, and there will be no issues. However, many issues arise if the objective is to show that the test treatment is similar to or not inferior to the active control, the so-called equivalence and noninferiority trial. These issues have been well recognized in the literature since the early 1980s. See, for example, Temple (1982), Temple (1997), Senn (2007), U.S. FDA (2010), and Rothmann, Wiens and Chan (2012). These issues will be discussed in Sections 2.5 through 2.7 of Chapter 2.

There is a consensus that use of a placebo is unethical and should be prohibited when an intervention shown to improve survival or decrease serious morbidity is available. See, for example, Temple and Ellenberg (2000),

Emanuel and Miller (2001), and Lyons (2001). Recognizing the inherent difficulties in assessing δ-equivalence/δ-no-worse-than, and hence the efficacy of the test treatment in an ACES, these authors also elaborate different scenarios where use of a placebo is ethical when there is an effective treatment, even though the Declaration of Helsinki (4th Revision 1996) is essentially excluding the use of placebo as a control in all clinical trials (Section 1.5.2). These difficulties arise in an ACES, but not in placebo-controlled trials (or superiority trials) as they relate to the assay sensitivity (ICH E10, 2001). See Sections 2.6 and 2.7 of Chapter 2 for further discussion. Note that the current version of the declaration (7th Revision; WMA 2013) allows use of placebo (or no treatment) when (1) there is no effective treatment or (2) under certain scenarios even there is an effective treatment (Section 1.5.2).

Sample size determination in the conventional null hypothesis testing in a placebo-controlled trial involves specifications of Type I (α) and Type II (β) error rates and δ_0. In practice, specification of δ_0 may be arbitrary—so are α and β to a lesser extent, although $\alpha = 0.05$ is typically used; as Feinstein (1975) wrote:

> What often happens is that the statistician and investigator decide on the size of δ_0. The magnitude of the sample is then chosen to fit the two requirements (1) that the selected number of patients can actually be obtained for the trial and (2) that their recruitment and investigation can be funded.

To specify δ_0, the paper continued: "In the absence of established standards, the clinical investigator picks what seems like reasonable value… If the sample size that emerges is unfeasible, δ_0 gets adjusted accordingly, and so do α and β, until n comes out right." Spiegelhalter and Freedman (1986) summarized precisely the practice in the specification of δ_0 in the following:

> There is very little explicit guidance as to how to set δ_0, and in practice it seems likely that δ_0 is juggled until it is set at a value that is reasonably plausible, and yet detectable given the available patients.

In the regulatory environment, however, an α level of one-sided 0.025 or two-sided 0.05 is the standard, and the study is typically powered at 80% or 90%. Actually, the estimate of the variability of the continuous endpoint or the background rate for the binary endpoint also plays a role in sample size calculation. Once the sample size is determined and the study is completed, δ_0 does not play any role in the statistical analyses or inferences. The null hypothesis of equality is either rejected or not. No inference may be made with regard to δ_0 when the null hypothesis is rejected, although point estimate and confidence interval provide information regarding the effect size.

It should be noted that too small δ_0 would result in a very large sample size, which is a waste of valuable resources. Furthermore, this may lead to an undesirable outcome where the treatment difference is statistically significant but is too small to be clinically meaningful. Very often δ_0 is set equal to twice (or more) of the minimum difference of clinical interest (Jones et al. 1996, p. 37).

On the other hand, an ACES is to show that the new treatment is sufficiently similar to (or not too much worse than) the standard therapy to be clinically indistinguishable (or noninferior). Therefore, the margin δ should be smaller than δ_0 (Jones et al. 1996; Kaul and Diamond 2007; ICH E9 1998; Liu 2000b; Suda et al. 2011). Jones et al (1996) suggests δ be no larger than half of δ_0, leading to sample sizes roughly four times as large as those in similar placebo-controlled trials.

1.6 Conventional Approach versus Role-Reversal Approach

In the 1970s, there was widespread recognition among statisticians that it is a flaw to accept the null hypothesis of no difference between two treatments (referred to as the conventional null hypothesis) when the null hypothesis is not rejected (e.g., Dunnett and Gent 1977; Makuch and Simon 1978). Consequently, many authors (e.g., Anderson and Hauck 1983; Blackwelder 1982) criticized the use of significance testing of the conventional null hypothesis (referred to here as "the conventional approach") in situations in which the experimenter wishes to establish the equivalence of two treatments. The main criticisms are that (1) two different treatments (or regimens) are not expected to have exactly the same treatment effect and (2) two treatments cannot be shown to be literally equivalent. Although the criticisms are legitimate, it is not practicable because (1) no other statistical methods can be used to establish the strict sense of equivalence (strict equality) and (2) the confidence interval approach and other forms of hypothesis testing (e.g., Anderson and Hauck 1983; Blackwelder 1982) (referred to here as "the role-reversal approach," e.g., two one-sided test procedure) were proposed to establish only δ-equivalence but not strict-sense equivalence (Ng 1995).

Can we use the conventional approach to establish that two treatments are δ-equivalent or that one treatment is δ-no-worse than the other treatment? The remainder of this section will address these questions.

When the conventional null hypothesis is not rejected, what can we conclude? Although accepting the conventional null hypothesis when it is not rejected is a flaw, failing to reject the conventional null hypothesis would lead one to believe that the treatment difference is not very "far" from zero. When the conventional null hypothesis is not rejected, one would believe that the larger the sample size, the closer to zero the treatment difference is. If the sample size is such that the Type II error rate (β error) at some δ is sufficiently small (e.g., < 0.025), then we can conclude that the two treatments are δ-equivalent. With this interpretation, Ng (1993a) showed that under the assumptions of normality with common known variance, the conventional approach to establish δ-equivalence and δ-no-worse-than coincides with the role-reversal approach if the Type II error rate (β error) in the conventional approach at δ (or $-\delta$) is equal to the Type I error rate (α error) in the role-reversal approach.

The argument for the equivalence of the two approaches for establishing δ-no-worse-than is briefly described as follows. It is obvious that the result holds if we are testing a simple null hypothesis against a simple alternative hypothesis, as opposed to a simple null against a composite alternative hypothesis. We then extend the simple alternative hypotheses in both approaches to the appropriate composite alternative hypotheses. Next, we perform the tests as if we are testing simple versus simple alternative hypotheses, except that in the conventional approach, we would conclude δ-no-worse-than instead of accepting the conventional null hypothesis if it is not rejected. The result then follows. The argument for the equivalence of the two approaches for establishing δ-equivalence is similar and is omitted here. Note that the argument for the equivalence of the two approaches assumes known variance rather than observed power calculated after the study is completed. Such a power calculation is not recommended (Hoenig and Heisey 2001).

Therefore, under the normality assumption with a common known variance, the conventional approach may be used to establish that two treatments are δ-equivalent or that one treatment is δ-no-worse than the other treatment. In practice, however, the conventional approach often cannot control the β error because the variance has to be estimated and the sample size might be smaller than planned due to dropouts, low recruitment, etc. On the other hand, the role-reversal approach does not have the difficulty in controlling the α error. That does not mean that the conventional approach is inappropriate in sample size calculation in the design of active-control equivalence studies. In fact, with proper choices of the α and β errors, the sample size calculation using either approach will give the same answer, where (1) the β error in the conventional approach is calculated at an alternative that the mean difference is δ and (2) the β error in the role-reversal approach is calculated at an alternative that the mean difference is zero (Ng 1995; Ng 1996).

The following discussions focus on δ-equivalence, although they are applicable to the δ-no-worse-than as well. It is incorrect to accept the conventional null hypothesis, regardless of the sample size. In fact, we can never establish the conventional null hypothesis in the strict sense no matter how large the sample size is, unless we "exhaust" the whole population (and hence, the true means are known). However, for fixed variance and fixed δ and with the same nonsignificant p-value, the larger the sample size, the stronger the evidence in supporting δ-equivalence. Furthermore, for fixed variance with the same nonsignificant p-value, the larger the sample size, the smaller the δ for claiming δ-equivalence (Ng 1995).

One should realize that establishing δ-equivalence has little or no meaning at all if δ is too large and that any two treatments are δ-equivalent if δ is large enough. For example, in antihypertensive studies for which the reduction in supine diastolic blood pressure is the primary efficacy variable, if δ = 8 mm Hg and the therapeutic effect of the standard therapy as compared to placebo is only 6 mm Hg, then we don't really gain anything by concluding that the test drug and the standard therapy are δ-equivalent because the placebo is also δ-equivalent to the standard therapy (Ng 1995).

1.7 Notations and Noninferiority Hypotheses

Throughout the rest of this book, T denotes the test or experimental treatment, S denotes the standard therapy or the active control, P denotes the placebo, and δ denotes the noninferiority (NI) margin. Furthermore, assume that there is no concurrent placebo control due to ethical reasons in life-threatening situations, for example. Note that T, S, and P could be the true mean responses for continuous outcomes (see Chapters 2 and 3) or the true success rates (or proportions of successes) for binary outcomes (see Chapter 4). Unless noted otherwise, we assume that a larger value corresponds to a better outcome.

The NI hypotheses ($H_0^{(1)}$ and $H_1^{(1)}$) in Section 1.2 may be restated using the notations introduced in this section as follows:

$$H_0: T - S \leq -\delta$$

versus

$$H_1: T - S > -\delta$$

or equivalently

$$H_0: T \leq S - \delta \qquad (1.3a)$$

versus

$$H_1: T > S - \delta \qquad (1.3b)$$

References

Anderson S, and Hauck WW (1983). A New Procedure for Testing Equivalence in Comparative Bioavailability and Other Clinical Trials. *Communications in Statistics: Theory and Methods*, **12**:2663–2692.

Blackwelder CW (1982). "Proving the Null Hypothesis" in Clinical Trials. *Controlled Clinical Trials*, **3**:345–353.

Blackwelder CW (1998). Equivalence Trials. In: Armitage P and Colton T eds. *Encyclopedia of Biostatistics*. New York: John Wiley, 1367–1372.

Chow SC (2011). Quantitative Evaluation of Bioequivalence/Biosimilarity. *J Bioequiv Availab* S1:002 doi: *http://omicsonline.org/0975-0851/JBB-S1-002.php*.

Chow S-C, and Liu J-P (2000). Individual Equivalence. In: Chow S-C ed. *Encyclopedia of Biopharmaceutical Statistics*. New York: Marcel Dekker, 259–266.

Chow S-C, and Liu J-P (2009). *Design and Analysis of Bioavailability and Bioequivalence Studies*, 3rd Edition. Boca Raton, FL: Chapman & Hall/CRC.

D'Agostino RB Sr, Massaro JM, and Sullivan LM (2003). Non-inferiority Trials: Design Concepts and Issues: The Encounters of Academic Consultants in Statistics. *Statistics in Medicine*, **22**:169–186.

Dunnett CW, and Gent M (1977). Significance Testing to Establish Equivalence Between Treatments, With Special Reference to Data in the form of 2 × 2 Tables. *Biometrics*, **33**:593–602.

Emanuel EJ, and Miller FG (2001). The Ethics of Placebo-Controlled Trials: A Middle Ground. *New England Journal of Medicine*, **345**:915–919.

European Agency for the Evaluation of Medicinal Products, Committee for Proprietary Medicinal Products (EMEA/CPMP, 1999). Concept Paper on Points to Consider: Choice of Delta. CPMP/EWP/2158/99. http://www.f-mri.org /upload/module-5/STAT_CHMP2158_Delta_choice.pdf (Accessed: September 8, 2013).

European Agency for the Evaluation of Medicinal Products, Committee for Proprietary Medicinal Products (EMEA/CPMP, 2000). Points to Consider on Switching Between Superiority and Noninferiority. http://www200B.ema. europa.eu/docs/en_GB/document_library/Scientific_guideline/2009/09 /WC500003658.pdf (Accessed: August 25, 2013).

European Agency for the Evaluation of Medicinal Products, Committee for Proprietary Medicinal Products (EMEA/CPMP, 2005). Guideline on the Choice of the Non-Inferiority Margin EMEA/CPMP/EWP/2158/99. http://www.ema.europa.eu /docs/en_GB/document_library/Scientific_guideline/2009/09 /WC500003636.pdf (Accessed: August 25, 2013).

European Agency for the Evaluation of Medicinal Products, Committee for Proprietary Medicinal Products (EMEA/CPMP, 2010). Guideline on Investigation of Bioequivalence. CHMP/EWP/QWP/1401/98 Rev. 1/Corr **, London, U.K.

Feinstein AR (1975). Clinical Biostatistics XXXIV. The Other Side of "Statistical Significance": Alpha, Beta, Delta, and the Calculation of Sample Size. *Clinical Pharmacology & Therapeutics*, **18**:491–505.

Fleming TR (2000). Design and Interpretation of Equivalence Trials. *American Heart Journal*, **139**:S171–176.

Gomberg-Maitland M, Frison L, and Halperin JL (2003). Active-Control Clinical Trials to Establish Equivalence or Noninferiority: Methodological and Statistical Concepts Linked to Quality. *American Heart Journal*, **146**(3):398–403.

Gülmezoglu AM, Widmer M, Merialdi1 M, et al. (2009). Active Management of the Third Stage of Labour Without Controlled Cord Traction: A Randomized Non-inferiority Controlled Trial. *Reproductive Health* (Open Access) http://www .reproductive-health-journal.com/content/6/1/2 (Accessed: March 30, 2013).

Hauck WW, and Anderson S (1999). Some Issues in the Design and Analysis of Equivalence Trials. *Drug Information Journal*, 33:109–118.

Hauschke D (2001). Choice of Delta: A Special Case. *Drug Information Journal*, **35**:875–879.

Hauschke D, Schall R, and Luus HG (2000). Statistical Significance. In: Chow S-C ed. *Encyclopedia of Biopharmaceutical Statistics*. New York: Marcel Dekker, 493–497.

Hauschke D, Steinijans V, and Pigeot I (2007). *Bioequivalence Studies in Drug Development: Methods and Applications*. Chichester, U.K.: John Wiley & Sons.

Hoenig JM, and Heisey DM (2001). The Abuse of Power: The Pervasive Fallacy of Power Calculations for Data Analysis. *The American Statistician*, **55**:19–24.

Hung H-MJ, Wang S-J, and O'Neill RT (2005). A Regulatory Perspective on Choice of Margin and Statistical Inference Issue in Non-inferiority Trials. *Biometrical Journal*, 47:28–36.

Hwang IK, and Morikawa T (1999). Design Issues in Non-inferiority/Equivalence Trials. *Drug Information Journal*, 33:1205–1218.

International Conference on Harmonization (ICH) E6 Guideline (1996). *Good Clinical Practice: Consolidated Guidance*. http://www.fda.gov/downloads/Drugs /GuidanceComplianceRegulatoryInformation/Guidances/UCM073122.pdf (Accessed: February 23, 2013).

International Conference on Harmonization (ICH) E9 Guideline (1998). *Statistical Principles for Clinical Trials*. http://www.fda.gov/downloads/Drugs /GuidanceComplianceRegulatoryInformation/Guidances/UCM073137.pdf (Accessed: September 27, 2012).

International Conference on Harmonization (ICH) E10 Guideline (2001). *Choice of Control Groups in Clinical Trials*. http://www.fda.gov/downloads/Drugs /GuidanceComplianceRegulatoryInformation/Guidances/UCM073139.pdf (Accessed: September 27, 2012).

Jones B, Jarvis P, Lewis JA, and Ebbutt AF (1996). Trials to Assess Equivalence: The Importance of Rigorous Methods. *British Medical Journal*, **313**:36–39.

Kaul S, and Diamond GA (2007). Making Sense of Noninferiority: A Clinical and Statistical Perspective on Its Application to Cardiovascular Clinical Trials. *Progress in Cardiovascular Diseases*, 49:284–299.

Lange S, and Freitag G (2005). Choice of Delta: Requirements and Reality: Results of a Systematic Review. *Biometrical Journal*, 47:12–27.

Lasagna L (1979). Editorial: Placebos and Controlled Trials Under Attack. *European Journal of Clinical Pharmacology*, 15:373–374.

Liu J-P (2000a). Equivalence Trials. In: Chow S-C ed. *Encyclopedia of Biopharmaceutical Statistics*. New York: Marcel Dekker, 188–194.

Liu J-P (2000b). Therapeutic Equivalence. In: Chow S-C ed. *Encyclopedia of Biopharmaceutical Statistics*. New York: Marcel Dekker, 515–520.

Lyons DJ (1999). Use and Abuse of Placebo in Clinical Trials. *Drug Information Journal*, 33:261–264.

Makuch R, and Johnson M (1990). Active Control Equivalence Studies: Planning and Interpretation. In: *Statistical Issues in Drug Research and Development*, K. Peace ed., New York: Marcel Dekker, 238–246.

Makuch R, and Simon R (1978). Sample size requirements for evaluating a conservative therapy. *Cancer Treatment Reports*, **62**:1037–1040.

Newell DJ (1992). Intention-to-Treat Analysis: Implications for Quantitative and Qualitative Research. *International Journal of Epidemiology*, **21**, 837–841.

Ng T-H (1993a). Broad-Sense Equivalence in an Active-Control Equivalence Study. Presented at the Third ICSA Applied Statistics Symposium, Plymouth Meeting, Pennsylvania; May 15, 1993.

Ng T-H (1993b). A Specification of Treatment Difference in the Design of Clinical Trials with Active Controls, *Drug Information Journal*, **27**:705–719.

Ng T-H (1995). Conventional Null Hypothesis Testing in Active-Control Equivalence Studies. *Controlled Clinical Trials*, **16**:356–358.

Ng T-H (1996). Letters to the Editor (A Reply to "A Note on Conventional Null Hypothesis Testing in Active-Control Equivalence Studies" by Hauschke D and Steinijans V). *Controlled Clinical Trials*, **17**:349–350.

Ng T-H (2001). Choice of Delta in Equivalence Testing. *Drug Information Journal*, **35**:1517–1527.

Ng T-H (2008). Noninferiority Hypotheses and Choice of Noninferiority Margin. *Statistics in Medicine*, **27**:5392–5406.

Piaggio G, Elbourne DR, Altman DG, Pocock SJ, and Evans SJ (2006). Reporting of Noninferiority and Equivalence Randomized Trials: An Extension of the CONSORT Statement. *Journal of American Medical Association*, **295**:1152–1160.

Pocock SJ (2003). Pros and Cons of Noninferiority Trials. *Blackwell Publishing Fundamental & Clinical Pharmacology*, **17**:483–490.

Rothmann MD, Wiens BL, and Chan ISF (2011). *Design and Analysis of Non-Inferiority Trials*. Boca Raton, FL: Chapman & Hall/CRC.

Schuirmann DJ (1987). A Comparison of the Two One-Sided Tests Procedure and the Power Approach for Assessing the Equivalence of Average Bioavailability. *Journal of Pharmacokinetics and Biopharmaceutics*, **15**:657–680.

Senn S (2007). *Statistical Issues in Drug Development*, 2nd edition. Chichester, U.K.: John Wiley & Sons.

Simon R (1999). Bayesian Design and Analysis of Active-Control Clinical Trials. *Biometrics*, **55**:484–487.

Simon R (2001). Therapeutic Equivalence Trials. In: Crowley J ed. *Handbook of Statistics in Clinical Oncology*, New York: Marcel Dekker, 173–187.

Spiegelhalter DJ, and Freedman LS (1986). A Predictive Approach to Selecting the Size of a Clinical Trial Based on Subjective Clinical Opinion. *Statistics in Medicine*, **5**:1–13.

Suda KJ, Hurley AM, McKibbin T, and Motl Moroney SE (2011). Publication of Noninferiority Clinical Trials: Changes over a 20-Year Interval. *Pharmacotherapy*, **31**(9):833–839.

Temple R, (1982). Government Viewpoint of Clinical Trials. *Drug Information Journal*, **16**:10–17.

Temple R (1997). When Are Clinical Trials of a Given Agent vs. Placebo No Longer Appropriate or Feasible? *Controlled Clinical Trials*, **18**:613–620.

Temple R, and Ellenberg SS (2000). Placebo-Controlled Trials and Active-Control Trials in the Evaluation of New Treatments. I. Ethical and Scientific Issues. *Annals of Internal Medicine*, **133**:455–463.

Treadwell J, Uhl S, Tipton K, Singh S, Santaguida L, Sun X, Berkman N, Viswanathan M, Coleman C, Shamliyan T, Wang S, Ramakrishnan R, and Elshaug A (2012). Assessing Equivalence and Noninferiority. Methods Research Report. (Prepared by the EPC Workgroup under Contract No. 290-2007-10063.) AHRQ Publication No. 12-EHC045-EF. Rockville, MD: Agency for Healthcare Research and Quality, June 2012. http://www.effectivehealthcare.ahrq.gov/ehc/products/365/1154/Assessing-Equivalence-and-Noninferiority_FinalReport_20120613.pdf (Accessed: March 1, 2014).

U.S. Food and Drug Administration (2001a). Guidance for Industry: Statistical Approaches to Establishing Bioequivalence, Center for Drug Evaluation and Research, U.S. Food and Drug Administration, Rockville, MD.

U.S. Food and Drug Administration (2001b). Guidance for Industry: Acceptance of Foreign Clinical Studies. http://www.fda.gov/RegulatoryInformation/Guidances/ucm124932.htm (Accessed: February 18, 2013).

U.S. Food and Drug Administration (2010). Draft Guidance for Industry: Non-inferiority Clinical Trials. http://.fda.gov/downloads/Drugs/Guidance Compliance RegulatoryInformation/Guidances/UCM202140.pdf (Accessed: August 25, 2013).

U.S. Food and Drug Administration (2012). Guidance for Industry and FDA Staff: FDA Acceptance of Foreign Clinical Studies Not Conducted Under an IND, Frequently Asked Questions. http://www.fda.gov/downloads/RegulatoryInformation/Guidances/UCM294729.pdf (Accessed: February 23, 2013).

Wellek S (2010). *Testing Statistical Hypotheses of Equivalence and Noninferiority*, 2nd edition. Boca Raton, FL: Chapman & Hall/CRC.

Wiens B (2002). Choosing an Equivalence Limit for Non-inferiority or Equivalence Studies. *Controlled Clinical Trials*, **23**, 2–14.

Wikipedia Contributors, Declaration of Helsinki. http://en.wikipedia.org/wiki/Declaration_of_Helsinki (Accessed: April 20, 2014).

Windeler J, and Trampisch H-J (1996). Recommendations Concerning Studies on Therapeutic Equivalence. *Drug Information Journal*, **30**:195–199.

World Medical Association (1989). Declaration of Helsinki: Ethical Principles for Medical Research Involving Human Subjects, 3rd Revision. http://history.nih.gov/research/downloads/helsinki.pdf (Accessed: February 24, 2013).

World Medical Association (1996). Declaration of Helsinki: Ethical Principles for Medical Research Involving Human Subjects, 4th Revision. http://www.birminghamcancer.nhs.uk/uploads/document_file/document/4f54bfed358e987415000194/Appx_III-VII.pdf (Accessed: February 24, 2013).

World Medical Association (2008). Declaration of Helsinki: Ethical Principles for Medical Research Involving Human Subjects, 6th Revision. http://www.wma.net/en/30publications/10policies/b3/17c.pdf (Accessed: February 24, 2013).

World Medical Association (2013). Declaration of Helsinki: Ethical Principles for Medical Research Involving Human Subjects, 7th Revision. *Journal of the American Medical Association*, 310(20):2191–2194. doi:10.1001/jama.2013.281053. http://jama.jamanetwork.com/article.aspx?articleid=1760318 (Accessed: April 8, 2014).

2

Choice of Noninferiority Margin for the Mean Difference

2.1 Introduction

This chapter elaborates the choice of the noninferiority (NI) margin as a small fraction (e.g., 0.2) of the therapeutic effect of the active control as compared to placebo based on the mean difference for a continuous endpoint.

The hypotheses stated in Section 1.7 of Chapter 1 can be shown graphically in Figure 2.1, where the axis is the true mean response, and the treatment gets better as we move to the right and worse if we move to the left. At the boundary, we have $T = S - \delta$. The null hypothesis is to the left of this boundary inclusively, and the alternative hypothesis is to the right of the boundary.

The International Conference on Harmonization (ICH) E9 document (1998) states that the "margin is the largest difference that can be judged as being clinically acceptable and should be smaller than differences observed in superiority trials of the active comparator." The ICH E10 document (2001) suggests that (1) the NI margin "cannot be greater than the *smallest effect size that the active drug would be reliably expected to have* compared with placebo in the setting of the planned trial"; and that (2) the margin "usually will be smaller than that suggested by the smallest expected effect size of the active control because of interest in ensuring that some clinically acceptable effect size (or fraction of the control drug effect) is maintained." Simon (2001) suggests that the NI margin δ must be no greater than the efficacy of S relative to P and will, in general, be a fraction of this quantity.

2.2 Proposed δ

How do you choose δ? This is a big question in testing for NI. To claim that the test treatment is not inferior to the standard therapy, we want δ to be small relative to the effect size of the active control, that is, $S - P$. Mathematically,

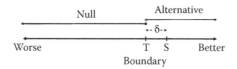

Alternative hypothesis: T > S − δ

Figure diagram:

Null   Alternative

← - δ - →

Worse   T   S   Better

Boundary

True mean response

FIGURE 2.1
Hypotheses.

δ should be some small ε between 0 and 1, such as 0.2 times the effect size, that is,

$$\delta = \varepsilon(S - P) \qquad (2.1)$$

At first sight, such a proposal appears to be useless because the effect size is not known. However, the effect size can and should be estimated from historical data (1) directly, where the active control was compared to the placebo in placebo-controlled trials, or (2) indirectly, where the active control was compared to another active control in a noninferiority trial. If no such data exists, the active control should not be used as a control in the NI trial because it is difficult to justify the NI margin without such data. For simplicity, we consider the situation where the effect size can and should be estimated directly from historical data, where the active control was compared to the placebo in placebo-controlled trials.

In February 2004, the European Agency for the Evaluation of Medicinal Products (EMEA) Committee for Proprietary Medicinal Products (CPMP) issued points to consider on the choice of NI margin for public comments (EMEA/CPMP 2004). It stated that "ideas such as choosing delta to be a percentage of the expected difference between active and placebo have been advocated, but this is not considered an acceptable justification for the choice." This is inconsistent with the EMEA/CPMP concept paper (1999)— see item 10 in Section 1.4 of Chapter 1. Seldrup (2004) made the following comments: "This may be too strong. In some situations it may be the 'right' (or only) choice meaningful to [the] statistician and clinician." However, no revision was made in response to these comments in the final document (EMEA/CPMP 2005).

The NI margin given by Equation 2.1 was first proposed by Ng (1993) and was motivated by regulatory considerations (see also Ng 2001). Although such a proposal does not answer the big question because the effect size is not known, it does reduce it to two smaller questions, which can be handled a lot easier than one big one (Ng 2008):

How do you choose ε?
How do you know the effect size?

For example, when answering the first question, we would have a better understanding of the rationale and interpretation of the NI margin when the effect size is assumed to be known. Furthermore, we could be more focused without worrying about different ways to estimate the effect size and other issues (e.g., assay sensitivity, constancy assumption, etc.). Issues with estimating the effect size will be discussed in Section 2.5. It makes sense to use the divide-and-conquer strategy.

2.3 Motivations

The proposed δ was first motivated by reviewing the study in a new drug application (NDA) submission in 1987 (Ng 1993, 2001). In this study, a new formulation and the old formulation of a cholesterol-lowering agent were compared with respect to their ability to maintain the depressed low-density lipoprotein (LDL) cholesterol level (see Figure 2.2). All patients were on the old formulation during the baseline run-in period of 2 weeks. Patients were then randomized to a 4-week treatment period with either the old formulation or the new formulation.

The LDL level was presumed to have been stabilized at baseline and was assumed to remain at the baseline level through the end of the experiment if patients continued on the old formulation. That would also be true for patients taking the new formulation, if the two formulations were equivalent. However, if patients were put on placebo, or if the new formulation was not effective, then the LDL level would be expected to eventually rise to the pre-baseline level. Therefore, to say that the two formulations are equivalent with respect to their ability to maintain the depressed LDL cholesterol level, we want the old formulation to be near the depressed level rather than near the pre-baseline level at the end of treatment period. Thus, δ should be a small fraction of the decrease in LDL level from pre-baseline to baseline. For NI trials, this means that δ should be a small fraction of the therapeutic effect of the standard therapy as compared to placebo (or the effect size).

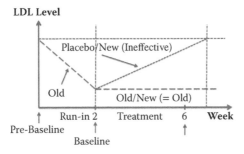

FIGURE 2.2
Comparison of two formulations of cholesterol-lowering agents..

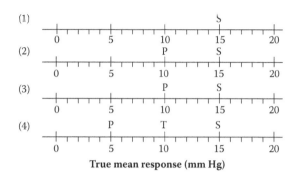

FIGURE 2.3
Four scenarios.

To further justify the proposal, let's consider the situation where we are comparing an antihypertensive agent versus a standard therapy (Ng 1993, 2001). Assume that there is a fictitious concurrent placebo control. The primary efficacy endpoint is the reduction in diastolic blood pressure.

Let's consider the four scenarios shown in Figure 2.3. In scenario 1, S is known, but P is not. In scenario 2, P is known, but S is not. Therefore, in the first two scenarios, the effect size, S − P, is unknown. In scenario 3, S − P = 5 and in scenario 4, S − P = 10. Is δ = 6 mm Hg acceptable? In the first two scenarios, we do not have enough information to answer the question. In scenario 3, δ = 6 mm Hg is unacceptable because claiming δ-equivalence of the test treatment to the standard therapy is meaningless, as placebo (P = 10) and the standard therapy (S = 15) would be δ-equivalent. In scenario 4, δ = 6 mm Hg is questionable because if T is halfway between placebo and the standard therapy (i.e., T = 10 mm Hg), then the test drug is δ-equivalent to both placebo and standard therapy. For the two treatments to be considered equivalent, the test drug should be closer to the standard therapy than to the placebo. For example, in scenario 4, δ should be less than 5 mm Hg. Therefore, δ should be less than one-half of the effect size. Or mathematically, $\delta = \varepsilon$ (S − P), where $0 < \varepsilon < 0.5$. This leads to the proposal that δ should be a small fraction of the effect size.

2.4 Choice of ε and Interpretation

How do you choose ε? In other words, what small fraction should be used? To answer this question, let's see how the proposed margin can be interpreted in terms of ε. Recall that we are testing the null hypothesis that $T \le S - \delta$, where $\delta = \varepsilon(S - P)$. At the boundary, we have

$$T = S - \delta = S - \varepsilon(S - P)$$

or equivalently,

$$T = P + (1-\varepsilon)(S-P)$$

as shown graphically in Figure 2.4, and we have the following interpretations:

1. If $\varepsilon = 0$, then $T = S$, so we would be testing for superiority against the standard therapy (i.e., H_0: $T = S$ vs. H_1: $T > S$).
2. If $\varepsilon = 1$, then $T = P$, so we would be testing for efficacy as compared to putative placebo (i.e., H_0: $T = P$ vs. H_1: $T > P$).
3. For $\varepsilon = 0.2$, it means that the test is 80% as effective as the standard therapy. We would conclude that the test treatment preserves greater than 80% of the control effect when the null hypothesis is rejected.

Thus, at the boundary, we put T on a scale of 0 to 1, with $\varepsilon = 0$, meaning that the test treatment is the same as the standard therapy (i.e., $T = S$), and with $\varepsilon = 1$, meaning that the test treatment is the same as placebo (i.e., $T = P$) (Ng 2001, 2008). To show NI, ε should be small, such as 0.2, but it is arbitrary.

Figure 2.5a shows a margin with $\varepsilon = 0.2$. If ε is increased to 0.5 (see Figure 2.5b), the boundary is halfway between the placebo and the standard therapy. When the null hypothesis is rejected, we cannot conclude NI because the test treatment could still be much worse than the standard therapy. However, we could conclude the efficacy of the test treatment because T > P when the null hypothesis is rejected, as shown in Figure 2.5b. This is also true if ε is further increased to, say, 0.8 as shown in Figure 2.5c. But we don't want ε to be so large that the test treatment could be worse than the placebo even if the null hypothesis is rejected. This could happen if ε is larger than 1 as shown in Figure 2.5d. Therefore, for efficacy, we want the NI margin to be no larger than the effect size or, mathematically, $\delta \leq S - P$; that is, $\varepsilon \leq 1$. This is essentially what ICH E9 and E10 suggested, but in a very fuzzy way, as they try to answer the big question rather than two smaller ones.

For example, if we know what the effect size is, then the margin must not be larger than the effect size; otherwise, we cannot conclude that the test treatment is better than placebo, even when the null hypothesis is rejected,

At the boundary:
$T = S - \varepsilon(S - P)$
T: Test; S: Standard therapy; P: Placebo

True mean response

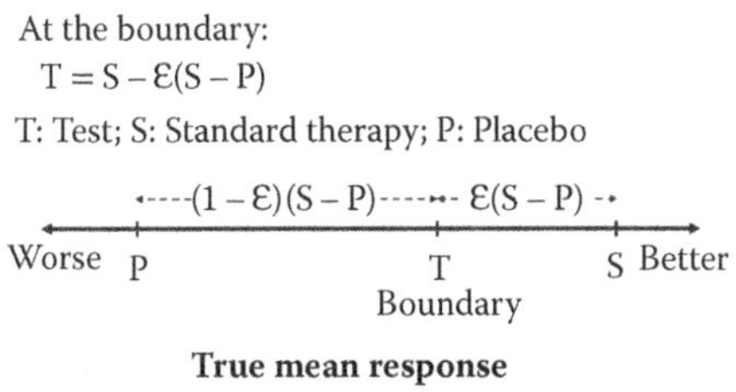

FIGURE 2.4
Interpretation.

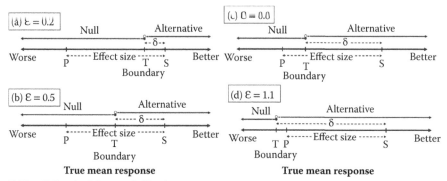

T: Test; S: Standard therapy; P: Placebo

Null hypothesis: T ≤ S – δ, δ = ε(S–P); Alternative hypothesis: T > S – δ

FIGURE 2.5
Choice of ε.

as shown in Figure 2.5d. On the other hand, if we set the margin to be no larger than the effect size, then we can conclude that the test treatment is better than placebo when the null hypothesis is rejected because T > S – δ ≥ S – (S – P) = P. However, the rationale given in the ICH E10 document is not clear, as the following demonstrates: "If this is done properly, a finding that the confidence interval for the difference between [the] new drug and the active control excludes a suitably chosen margin could provide assurance that the drug has an effect greater than zero."

Tsong et al. (2003) discussed three objectives of NI trials: (1) establishing evidence of efficacy over placebo, (2) preserving a specific percentage of the effect size of the active control, and (3) demonstrating the test treatment being "not much inferior" to the active control. After fairly lengthy derivations, the hypotheses for all three objectives can be formulated as

$$H_0: \mu_T - \mu_A = (\lambda - 1)(\mu_A - \mu_P) \text{ versus } H_1: \mu_T - \mu_A > (\lambda - 1)(\mu_A - \mu_P)$$

where $0 \leq \lambda < 1$, or equivalently,

$$H_0: T - S = (\lambda - 1)(S - P) \text{ versus } H_1: T - S > (\lambda - 1)(S - P)$$

using the notations in Section 1.7 of Chapter 1. These hypotheses are essentially the same as Equations 1.3a and 1.3b with

$$\varepsilon = (1 - \lambda)$$

Therefore, the term (1 – λ) in Tsong et al. (2003) corresponds to the ε in this book. Testing for preservation or retention of, say, 80% of the active-control effect is boiled down to testing for NI, with the NI margin given by Equation 2.1, with $\varepsilon = 0.2$ (see also Hung, Wang, and O'Neill 2005). It should be noted that when the null hypothesis is rejected with $\varepsilon = 0.2$, we would conclude that the test treatment preserves greater than 80% of the control effect, rather than at least 80% of the control effect.

In many situations, a small ε would lead to a sample size that is impractical. In that case, we cannot show NI. What we can do is use a larger δ and show the efficacy of the test treatment as compared to placebo. For example, in 1992, the U.S. Food and Drug Administration (FDA) Cardio-Renal Advisory Committee recommended the use of half of the effect size of the standard therapy as the NI margin for trials of new thrombolytics (see Chapter 11 for details). This recommendation is reasonable if the objective is to establish the efficacy of the test product (as compared to placebo) without claiming NI. For the latter objective, however, a much smaller δ should be used. We could also claim preservation of more than 50% of the effect when the null hypothesis is rejected if the constancy assumption holds. The constancy assumption will be discussed in Section 2.5.1.

Therefore, the choice of ε depends on the study objective. If the objective is to establish the efficacy of the test treatment as compared to placebo through its comparison to the standard therapy, then ε can be as large as 1 (Ng 1993, 2001, 2008). However, to claim equivalence or NI, ε should be small. How small is small? Perhaps ε should depend on other benefits (e.g., better safety profiles) relative to the primary endpoint. However, it is not clear how to determine ε based on this. Such a determination could be very subjective.

2.5 How Do You Know the Effect Size?

How do you know the effect size? This question is much more difficult to answer than the first one (see Section 2.2), but it is a lot easier to answer than the original question. A short answer: We don't know the effect size. A long answer: It must be estimated from historical data, which leads to a controversial issue with regard to the constancy assumption (see Section 2.5.1). For simplicity, assume that there is only one prior study of the active control compared to a placebo. Chapter 7 will discuss situations where there are multiple studies.

2.5.1 Constancy Assumption

The constancy assumption says that the effect size in the current NI trial is equal to the effect size in the historical trials. This assumption allows us to estimate the effect size in the current trial using the historical data. Mathematically, the constancy assumption means

$$S - P = (S - P)_h$$

where $(S - P)_h \equiv S_h - P_h$ denotes the effect size in the historical trials and S_h and P_h denote the mean responses for the standard therapy and placebo in the historical trials, respectively.

Constancy assumption depends on three major factors: (1) time difference between the historical and current trials, (2) study design, and (3) trial conduct:

1. If the historical trial was conducted, say, 10 years ago, we have to worry about changes over time, such as medical practices and patient population.

2. As discussed in the ICH E10 document (2001), the current study design should be similar to that of previous studies. For example, the current study should enroll a similar patient population, using the same inclusion/exclusion criteria. The same dosage regimen and the same efficacy endpoint should be used with the same duration/follow-up, etc. In this regard, the study design should be adequately reported. See Schulz, Altman, and Moher (2010) and Piaggio et al. (2012) for guidelines on reporting parallel group randomized trials.

3. The conduct of the current study also has an impact on the constancy assumption. For example, mixing up treatment assignments would bias toward similarity of the two treatments and reduce the effect size of the active control when the test treatment is not effective or is less effective than the active control (Ng 2001).

Setting aside the conduct of the study and assuming that the prior study was conducted recently, it is reasonable to believe that the constancy assumption would hold if the current study follows the same protocol with the same centers/investigators as a recent placebo control study of the active control. In fact, it is logical to expect both studies to have the same (true) mean response for the active control (as well as for the placebo if the placebo were also included in the current study), although in reality, the results may show otherwise, most likely due to the conduct of the study. Such an expectation implies the constancy assumption (i.e., $S = S_h$ and $P = P_h$, which implies $S - P = S_h - P_h$), but not vice versa. In other words, the constancy assumption is weaker than the assumption that the active control and placebo have the same (true) mean response in both studies. Changing the protocol in the current study from that of the placebo control study of the active control would likely invalidate the later assumption, while the constancy assumption may still hold. However, it is difficult to assess the impact of such changes on the constancy assumption.

2.5.2 Discounting

One way to alleviate concern that the constancy assumption is being violated is to discount the effect size estimated from the historical data as discussed by Ng (2001). Mathematically, we let $S - P = \gamma(S - P)_h$, where $0 < \gamma \le 1$ is the discount factor. This discount factor may be divided into two components (Ng 2008). The first component is an adjustment for the changes over time. The second component discounts the effect size due to the potential bias as a result of differences in the study design and conduct of the trial. The second component may be further divided into two subcomponents, one due to the study design and the other one due to the conduct of the trial. Thus, γ may be divided into three components, say, γ_1, γ_2, and γ_3, corresponding to the three major factors given in Section 2.5.1. Again, the divide-and-conquer strategy is used here.

How much discounting (i.e., γ) should be used in practice? It is difficult, if not impossible, to justify the choice of γ statistically and/or scientifically. In fact, choosing γ is more difficult than choosing ε because medical practices and patient populations may change over time, and γ depends on the design and conduct of the NI trial, as noted in Section 2.5.1. Unfortunately, one may have to rely heavily on subjective judgment in choosing γ. Incorporating the discounting factor γ into the NI margin defined by Equation 2.1, we have

$$\delta = \varepsilon\gamma(S-P)_h \qquad (2.2)$$

Since γ_3 depends on the conduct of the NI trial, the constancy assumption cannot be fully assessed before the study is completed. At the design stage, one approach is to assume that the NI trial will be properly conducted and hope for the best. A conservative approach is to build in a "safety" margin (e.g., use $\gamma_3 = 0.95$) just in case, but at the expense of a larger sample size.

2.6 Assay Sensitivity

"Assay sensitivity is a property of a clinical trial defined as the ability to distinguish an effective treatment from a less effective or ineffective treatment" (ICH E10 2001, p. 9). For an NI clinical trial, assay sensitivity is the trial's ability to detect a difference between treatments of a specified size M_1, defined as the entire assumed treatment effect of the active control in the NI trial (U.S. FDA 2010).

An NI trial without assay sensitivity is uninterpretable because (1) a finding of NI in such a trial does not rule out the possibility that the test treatment is a lot less effective than the standard therapy by the definition of assay sensitivity, and (2) the efficacy of the experimental treatment cannot be inferred. Therefore, it is essential that an NI trial have assay sensitivity. The presence of assay sensitivity in a noninferiority/equivalence trial may be deduced from two determinations, as follows (ICH E10 2001):

1. Historical evidence of sensitivity to drug effects (HESDE), i.e., that similarly designed trials in the past regularly distinguished effective treatments from less effective or ineffective treatments.

2. Appropriate trial conduct, i.e., that the conduct of the trial did not undermine its ability to distinguish effective treatments from less effective or ineffective treatments.

Appropriateness of trial conduct can only be fully evaluated after the active control noninferiority trial is completed. Not only should the design of the noninferiority trial be similar to that of previous trials used to determine historical evidence of sensitivity to drug effects (e.g., entry criteria, allowable concomitant therapy); but, in addition, the actual study population entered, the concomitant therapies actually used, etc., should be assessed to ensure that conduct of the study was, in fact, similar to the previous trials. The trial should also be conducted with high quality (e.g., good compliance, few losses to follow-up). Together with historical evidence of sensitivity to drug effects, appropriate trial conduct provides assurance of assay sensitivity in the new active-control trial.

2.7 Constancy Assumption versus Assay Sensitivity

Although similar considerations, such as the study design and conduct of the trials, are suggested in the assessment of the constancy assumption (Section 2.5.1) and assay sensitivity (Section 2.6), they are two different concepts. The former is an assumption regarding the unknown parameters (the true effect sizes), which cannot be verified and does not depend on the sample size. The latter is a property of a clinical trial and does depend on the sample size. For example, if the constancy assumption holds, it does not necessarily mean the trial has assay sensitivity because the trial size may be too small. On the other hand, if the constancy assumption does not hold, the trial may still have assay sensitivity if the trial size is large enough. In a poorly conducted trial, it is most likely that the constancy assumption does not hold and that the trial does not have assay sensitivity. From a statistical point of view, a well-conducted NI trial with adequate power to detect a treatment difference as small as the NI margin will have assay sensitivity.

2.8 Discussion

As discussed in Section 2.3, to show NI, ε should be no larger than 0.5. A treatment difference larger than 50% of the effect size is probably clinically important. However, it is not clear how much smaller it should be and still be considered clinically important. This corresponds to the top-down approach discussed in Section 1.4 of Chapter 1. On the other hand, a very small fraction (e.g., 0.0001) of the effect size is probably clinically acceptable. However, it is not clear how much larger it should be and still be considered clinically acceptable. This corresponds to the bottom-up approach, which was discussed in Section 1.4 of Chapter 1. Gülmezoglu et al. (2009) (see Item 14 in Section 1.4 of Chapter 1) suggested $\varepsilon = 0.2$, partially in consideration of the sample size.

Snapinn (2004) discussed two ways of discounting historical data to make inferences about the efficacy of the test treatment relative to placebo: (1) Using the fixed-margin approach discussed in Section 5.3 of Chapter 5 as a way of discounting for its inefficiency in utilizing the data, and (2) preserving a fraction of the active control's effect. Although discounting may be viewed as an ineffective method, there is no quantitative measure for such discounting (Ng 2001). Hung, Wang, and O'Neill (2005) discussed the use of retention (or preservation) of the control effect as a cushion in assessing the efficacy of the experimental treatment in the event that the constancy assumption may be violated (see also Wang and Hung 2003).

Using preservation as a form of discounting would be confusing and may lead to inappropriate "double credit" (see Section 5.7 of Chapter 5). An invalid constancy assumption should be dealt with by discounting rather than by preservation, although it is mathematically correct (Ng 2008). The divide-and-conquer strategy makes it clear that preservation and discounting are two different concepts, although they are indistinguishable mathematically (Ng 2001, 2008). See Section 5.7 of Chapter 5 for a further discussion on these concepts.

References

European Agency for the Evaluation of Medicinal Products, Committee for Proprietary Medicinal Products (EMEA/CPMP, 1999). Concept Paper on Points to Consider: Choice of Delta. CPMP/EWP/2158/99. http://www.f-mri.org/upload/module-5/STAT_CHMP2158_Delta_choice.pdf (Accessed: September 8, 2013).

European Agency for the Evaluation of Medicinal Products, Committee for Proprietary Medicinal Products (EMEA/CPMP, 2004). Points to Consider on the Choice of Non-Inferiority Margin (Draft). CPMP/EWP/2158/9.

European Medicines Agency (2005). Committee for Medicinal Products for Human Use (CHMP). Guideline on the Choice of the Non-Inferiority Margin EMEA/CPMP/EWP/2158/99.

Gülmezoglu AM, Widmer M, Merialdi M, et al. (2009). Active Management of the Third Stage of Labour Without Controlled Cord Traction: A Randomized Noninferiority Controlled Trial. *Reproductive Health* (Open Access) http://www.reproductive-health-journal.com/content/6/1/2 (Accessed: March 30, 2013).

Hung HMJ, Wang S-J, and O'Neill RT (2005). A Regulatory Perspective on Choice of Margin and Statistical Inference Issue in Non-Inferiority Trials. *Biometrical Journal*, **47**:28–36.

International Conference on Harmonization (ICH) E9 Guideline (1990). *Statistical Principles for Clinical Trials.* http://www.fda.gov/downloads/Drugs /GuidanceComplianceRegulatoryInformation/Guidances/UCM073137.pdf (Accessed: September 27, 2012).

International Conference on Harmonization (ICH) E10 Guideline (2001). *Choice of Control Groups in Clinical Trials.* http://www.fda.gov/downloads/Drugs /GuidanceComplianceRegulatoryInformation/Guidances/UCM073139.pdf (Accessed: September 27, 2012).

Ng T-H (1993). A Specification of Treatment Difference in the Design of Clinical Trials with Active Controls, *Drug Information Journal*, **27**:705–719.

Ng T-H (2001). Choice of Delta in Equivalence Testing, *Drug Information Journal*, **35**:1517–1527.

Ng T-H (2008). Noninferiority Hypotheses and Choice of Noninferiority Margin, *Statistics in Medicine*, **27**:5392–5406.

Piaggio G, Elbourne DR, Pocock SJ, Evans SJW, and Altman DG (2012). Reporting of Noninferiority and Equivalence Randomized Trials: Extension of the CONSORT 2010 Statement. *Journal of the American Medical Association*, **308**(24):2594–2604.

Schulz KF, Altman DG, and Moher D (2010). CONSORT 2010 Statement: Updated Guidelines for Reporting Parallel Group Randomised Trials, *British Medical Journal*, **340**:698–702.

Seldrup J (2004). Comments Regarding CPMP/EWP/2158/99 Draft: Points to Consider on the Choice of Non-Inferiority Margin. *International Society for Clinical Biostatistics News*, June 2004.

Simon R (2001). Therapeutic Equivalence Trials. In: Crowley J ed. *Handbook of Statistics in Clinical Oncology*, New York: Marcel, 173–187.

Snapinn SM (2004). Alternatives for Discounting in the Analysis of Noninferiority Trials. *Journal of Biopharmaceutical Statistics*, **14**:263–273.

Tsong Y, Wang S-J, Hung H-MJ, and Cui L (2003). Statistical Issues on Objectives, Designs and Analysis of Non-inferiority Test Active Controlled Clinical Trials. *Journal of Biopharmaceutical Statistics*, **13**:29–41.

U.S. Food and Drug Administration (2010). Draft Guidance for Industry: Non-inferiority Clinical Trials. http://www.fda.gov/downloads/Drugs /GuidanceComplianceRegulatoryInformation/Guidances/UCM202140.pdf (Accessed: August 25, 2013).

Wang S-J and Hung H-MJ (2003). Assessment of Treatment Efficacy in Non-inferiority Trials. *Controlled Clinical Trials*, **24**:147–155.

3

Choice of Noninferiority Margin for the Mean Ratio and Hazard Ratio

3.1 Introduction

In Chapter 2, the hypotheses for a continuous endpoint are stated in terms of the mean difference rather than the mean ratio, and the endpoint can be negative, such as a change from baseline. When the outcome measurement is positive, hypothesis testing based on the mean ratio could be more desirable. One such example is the bioequivalence study where we want to show that the mean ratio of the test product over the reference product is between 0.8 and 1.25 based on the area under the concentration curve (AUC) (Schuirmann 1987) (see also Section 1.1 of Chapter 1). Furthermore, for survival or time to event (e.g., time to progression) endpoints, the hazard ratio is commonly used for comparing treatments (see, e.g., Rothmann et al. 2003).

3.2 Why Mean Ratio?

To show that hypothesis testing based on the mean difference could be undesirable when the outcome measurement is positive and could vary widely, let us consider the two-dimensional space shown in Figure 3.1, where the horizontal axis is the mean response S and the vertical axis is the mean response T. The 45-degree dashed line passing through the origin is given by $T = S$. Regardless of the value of S, the null hypothesis based on the mean difference is below the parallel line given by $T = S - \delta$, inclusively, and the alternative hypothesis is above that parallel line. The margin δ does not depend on S, since we are testing the hypothesis in terms of the mean difference. This could be undesirable when the response measurement is strictly positive and S could vary widely. For example, a fixed-margin δ could be small relative to S and is acceptable if S is large, but the margin δ could be too large if S is small, as shown in Figure 3.1. Furthermore, "101 versus 100" appears to be

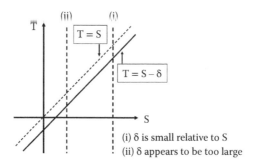

FIGURE 3.1
Mean difference.

different from "2 versus 1," although the differences in both cases are equal. For those reasons, we might want to consider testing noninferiority (NI) in terms of the mean ratio rather than the mean difference.

3.3 Mean Ratio for Continuous Endpoints

The hypotheses in terms of the mean ratio are more difficult conceptually. For the mean ratio, we test the null hypothesis that

$$T/S \leq 1/r \tag{3.1}$$

against the alternative hypothesis that

$$T/S > 1/r$$

where r ($\geq$ 1) acts like the NI margin, although it does not look like one. If r = 1, then we would test for superiority against the standard therapy. On the other hand, if r = S/P > 1 (assuming S > P), we would test for efficacy as compared to putative placebo and S/P acts like the effect size. If we knew this effect size, then a natural way to determine r is to take this effect size and raise it to a power of ε; that is, $r = (S/P)^\varepsilon$, for $0 \leq \varepsilon \leq 1$.

Although it is not intuitively clear if the proposed r makes sense, everything does become clear after taking a log transformation. More specifically, we would test the null hypothesis that

$$\log(T) - \log(S) \leq -\log(r)$$

against the alternative hypothesis that

$$\log(T) - \log(S) > -\log(r)$$

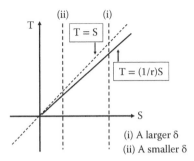

FIGURE 3.2
Mean ratio.

where $\log(r) = \varepsilon\ [\log(S) - \log(P)]$. Note that the treatment difference is measured on a log scale, as is the effect size. Again, the NI margin is equal to ε times the effect size on a log scale.

Note that the null hypothesis given by Equation 3.1 can be rewritten as

$$T - S \leq -\delta$$

where $\delta = (1 - 1/r)S$, so the NI margin depends on S, as shown in Figure 3.2.

However, when testing the null hypothesis given by Equation 3.1 with a fixed constant r, it is not advisable to compare the lower confidence limit for $(T - S)$ with $-\delta$ computed using the point estimate of S, because the variability of estimating S has not been taken into consideration in such a comparison; more importantly, S appears in both sides of the inequality, and the margin depends on S. Instead, the null hypothesis given by Equation 3.1 could be tested by comparing the lower confidence limit for T/S (with or without a log transformation) or $T - (1/r)S$ with the appropriate critical values. See Laster and Johnson (2003) for further discussion on testing the null hypothesis given by Equation 3.1 for a fixed r.

3.4 Hazard Ratio for Survival Endpoints

For time-to-event endpoints, let T, S, and P denote the hazards for the test treatment, the standard therapy, and the placebo, respectively. Note that a smaller hazard corresponds to a better outcome. This is different from the assumption for continuous data, where we assume that a larger value corresponds to a better outcome. Therefore, the NI hypothesis based on the hazard ratio looks at the opposite direction from the NI hypothesis based on the mean ratio. Accordingly, we test the null hypothesis that

$$T/S \geq r \tag{3.2a}$$

against the alternative hypothesis that

$$T/S < r \qquad\qquad (3.2b)$$

where $r \geq 1$ acts like the NI margin, although it does not look like one. Following the same ideas for the mean ratio in Section 3.3, we set

$$r = (P/S)^{\varepsilon}$$

for $0 \leq \varepsilon \leq 1$, if we know the effect size P/S (assuming $S < P$). Taking a log transformation, we would test the null hypothesis that

$$\log(T) - \log(S) \geq \log(r)$$

against the alternative hypothesis that

$$\log(T) - \log(S) < \log(r) \qquad\qquad (3.3)$$

where $\log(r) = \varepsilon\,[\log(P) - \log(S)]$. Note that the NI margin is equal to ε times the effect size on a log scale.

Rothmann et al. (2003) gave the arithmetic and geometric definitions of the proportion of active-control effect retained, recognizing that the latter is more appropriate than the former. They proposed to test whether the treatment maintains $100\delta_0$ percent of the active control, where $0 < \delta_0 < 1$. With the geometric definition, the alternative hypothesis given by (1b) of Rothmann et al. (2003), that is, $H_1: \log HR(T/C) < (1 - \delta_0)\log HR(P/C)$, is essentially the same as the alternative hypothesis given by Equation 3.3 with $\varepsilon = 1 - \delta_0$.

If one is interested in survival at a fixed time point, such as in the thrombolytic area where the primary endpoint is 30-day mortality (see Chapter 11), then the outcome variable becomes binary (see Chapter 4). In that case, the ratio is better known as the relative risk (RR), and the NI hypotheses can be set up similarly to those for the hazard ratio as given by Equations 3.2a and b.

3.5 Discussion

The NI hypotheses based on the mean difference (D) and the mean ratio (R) for a continuous outcome variable are characterized by Figures 3.1 and 3.2, respectively. Expressing R in the form of D with $\delta = (1 - 1/r)S$ (see Section 3.3), or vice versa with $r = 1/(1 - \delta/S)$, would not change its original characteristic because δ and r depend on S. Note that the proposed margins of $\delta = \varepsilon(S - P)$ and $r = (S/P)^\varepsilon$ also depend on S, but only through $(S - P)$ and S/P, respectively.

In placebo-controlled trials of the active treatment (or in superiority trials in general) with a continuous outcome variable, the efficacy is usually measured by the mean difference, not the mean ratio. However, if the analysis is performed using log-transformed data (this implicitly assumes that the outcome variable is positive), the efficacy is actually measured by the mean ratio. The statistical test based on the mean difference (T_d) may or may not agree in general with the statistical tests based on the mean ratio (T_r) due to the following reasons. Although adding a constant to all data points will not affect the test based on the mean difference (e.g., t-test), doing so with a positive (negative) constant will result in (1) a smaller (larger) mean ratio if the mean ratio is larger than 1 and (2) a larger (smaller) mean ratio if the mean ratio is less than 1.

Although the choice of D or R should be determined by what was used in the historical studies of active-control treatment, the type of the continuous outcome variable should also be taken into consideration. More specifically, if the outcome variable could take a negative value (e.g., change from baseline), then D is the only choice. D or R may be considered in a situation with a positive outcome variable. Furthermore, if S could vary greatly, then R is preferable, as discussed in Section 3.2. Such considerations are also applicable in the design of placebo-controlled trials, so that the NI trial would have the same metric as the placebo-controlled trials of the active-control treatment. There are practical problems in the design of an NI trial if T_d was used when T_r should have been used in the placebo-controlled trials of the active-control treatment. For example, T_r might not be statistically significant, or the raw data might not be available for estimating the effect size based on the mean ratio in the determination of the NI margin.

Reasons for testing for NI based on mean ratio rather than mean difference are discussed in Section 3.2. Specifying the NI margin as a fraction of the active control (such as, ε_0, where $0 < \varepsilon_0 < 1$) would result in testing for NI based on the mean ratio because $T - S \le -\delta \equiv -\varepsilon_0 \cdot S$ if and only if $T/S \le (1 - \varepsilon_0)$ (Laster and Johnson 2003; Hshieh and Ng 2007). Note that (1) choosing NI margin as a fraction of the active control would not work if the outcome variable may take both negative and positive values, and (2) there is no such thing as placebo in the example discussed by Hshieh and Ng (2007).

The "retention hypothesis" (see, e.g., Hung, Wang, and O'Neill 2005) is, in fact, based on the mean ratio after "taking away" the placebo effect. If there is no placebo effect (i.e., $P = 0$), it reduces to testing the mean ratio (Hauschke 2001). Note that the "retention hypothesis" is the same as the NI hypothesis given by Equation 1.3a, with the NI margin given by Equation 2.1 with percent retention equal to $(1 - \varepsilon)$.

References

Hauschke D (2001). Choice of Delta: A Special Case. *Drug Information Journal*, **35**:875–879.

Hshieh P and Ng T-H (2007). Noninferiority Testing with a Given Percentage of the Control as the Noninferiority Margin. *Proceedings of the American Statistical Association*, Biopharmaceutical Section [CD-ROM], Alexandria, VA: American Statistical Association. Note: It appeared in the Health Policy Statistics section by mistake.

Hung H-MJ, Wang S-J, and O'Neill RT (2005). A Regulatory Perspective on Choice of Margin and Statistical Inference Issue in Non-Inferiority Trials. *Biometrical Journal*, **47**:28–36.

Laster LL and Johnson MF (2003). Non-inferiority Trials: The "At Least as Good As" Criterion. *Statistics in Medicine*, **22**:187–200.

Rothmann M, Li N, Chen G, Chi GY-H, Temple R, and Tsou H-H. (2003). Design and Analysis of Non-Inferiority Mortality Trials in Oncology. *Statistics in Medicine*, **22**:239–264.

Schuirmann DJ (1987). A Comparison of the Two One-Sided Tests Procedure and the Power Approach for Assessing the Equivalence of Average Bioavailability. *Journal of Pharmacokinetics and Biopharmaceutics*, **15**:657–680.

4

Noninferiority Hypotheses with Binary Endpoints

4.1 Introduction

Let's consider a binary endpoint where an outcome could be a success or a failure. The same notations will be used—that is, T, S, and P—to denote the true success rates for the test treatment, the standard therapy, and the placebo, respectively.

4.2 Noninferiority Hypothesis Based on the Difference or Ratio of Two Proportions

For binary data, using $\delta = \varepsilon(S - P)$ as the noninferiority (NI) margin for testing NI based on the difference of two proportions could be problematic when the success rate for the standard therapy is very high. For example, if $S = 0.95$, $P = 0.45$, and $\varepsilon = 0.2$, then $\delta = 0.1$. So, we test the null hypothesis that

$$T - S \leq -0.1$$

against the alternative hypothesis that

$$T - S > -0.1$$

When the null hypothesis is rejected, we would conclude that

$$T > 0.95 - 0.1 = 0.85$$

This appears to be reasonable at first based on the success rate. However, the failure rate for the standard therapy is 0.05, while we conclude that the failure rate for the test treatment is less than 0.15. It is questionable to claim NI when we could not rule out doubling of the failure rate.

Similarly, using $r = (S/P)^\varepsilon$ as the NI margin (see Section 3.3 of Chapter 3) for testing NI based on the ratio of two proportions could be problematic when the success rate for the standard therapy is very high. For example, if $S = 0.95$, $P = 0.45$, and $\varepsilon = 0.2$, then $r = 1.1612$. So, we test the null hypothesis that

$$T/S \leq 1/1.1612$$

against the alternative hypothesis that

$$T/S > 1/1.1612$$

When the null hypothesis is rejected, we would conclude that

$$T > S/1.1612 = 0.8181$$

Again, it is questionable to claim NI when we could not rule out doubling of the failure rate. Due to these potential problems, the NI hypothesis based on the odds ratio is proposed next.

4.3 NI Hypothesis Based on the Odds Ratio

Let $O(X) = X/(1 - X)$ denote the odds, where $X = T, S$, or P, and test the null hypothesis that

$$O(T)/O(S) \leq 1/r \tag{4.1a}$$

against the alternative hypothesis that

$$O(T)/O(S) > 1/r \tag{4.1b}$$

where $r \geq 1$ acts like the NI margin although it does not look like one. Following the same idea for the mean ratio in Section 3.3 of Chapter 3, we set

$$r = [O(S)/O(P)]^\varepsilon$$

for $0 \leq \varepsilon \leq 1$ if we know the effect size $O(S)/O(P)$ (assuming $S > P$). Taking a log transformation, we would test the null hypothesis that

$$\log[O(T)] - \log[O(S)] \leq -\log(r)$$

against the alternative hypothesis that

$$\log[O(T)] - \log[O(S)] > -\log(r)$$

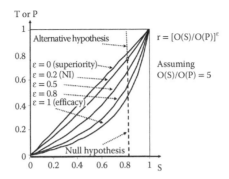

FIGURE 4.1
Odds ratio.

where $\log(r) = \varepsilon \, [\log[O(S)] - \log[O(P)]]$. Again, the NI margin is equal to ε times the effect size on a log scale.

The hypotheses based the odds ratio are shown graphically in Figure 4.1 using the original scale (i.e., success rate) with $\varepsilon = 0, 0.2, 0.5, 0.8$, and 1, where we assume that $O(S)/O(P) = 5$. For example, with $\varepsilon = 0.2$, the null hypothesis for any given S is given by the dashed line, and the alternative hypothesis is given by the dotted line. Note that the NI margin "adjusts" very smoothly when S ranges from 0 to 1.

Continuing the example in Section 4.2 where S = 0.95, P = 0.45, and $\varepsilon = 0.2$, it can be shown that a rejection of the null hypothesis based on the odds ratio implies that T > 0.9101, which guards against a doubling of the failure rate. Note that the effect size of 23.22 is very large as measured by the odds ratio [i.e., $O(S)/O(P)$].

4.4 Discussion

The concepts of (1) "x% as effective (or good) as" (Ng 1993, 2001; Simon 1999), (2) the preservation of the effect of the standard therapy (Tsong et al. 2003) (see Section 2.4 of Chapter 2), and (3) the retention of the active-control effect (Rothmann et al. 2003) (see Section 3.4 of Chapter 3) are essentially the same. These concepts come naturally when the NI margin is determined as the small fraction of the therapeutic effect of the active control as compared to placebo as discussed in Chapter 2, although a potential problem arises when the proposed NI margin is applied to the difference in proportions as discussed in Section 4.2. A similar problem is also discussed by Hung, Wang, and O'Neill (2005), where the "arithmetic" (in the sense of Rothmann et al. 2003) (see Section 3.4 of Chapter 3) version of control effect retention is used in assessing the relative risk. Such a problem could be avoided by using the "geometric" version of control effect retention, which is defined on the log scale as discussed in Section 3.3 of Chapter 3.

Wellek (2005) discussed three measures of dissimilarity for binary endpoints, namely, the odds ratio, the relative risk, and the difference. The author prefers the odds ratio over the other two measures, partly because the NI hypotheses defined in terms of the other two measures are bounded by lines that cross the boundaries of the parameter space. Garrette (2003) argued that the odds ratio is the most rational measure for assessing therapeutic equivalence and NI for binary outcomes and that there are clear advantages to expressing margins in terms of the odds ratio. Section 4.2 further enhances the reasons for using the odds ratio. In addition, using the logit link function in the framework of a generalized linear model (e.g., McCullagh and Nelder 1990, 31) would result in an analysis based on the log odds ratio.

Garrett (2003) proposed an odds ratio lower margin of 0.5, while Tu (1998) and Senn (2000) suggested values of 0.43 and 0.55, respectively. Such margins, however, do not take the placebo success rate (i.e., P using the notation in this book) into consideration, and may be too "low" depending upon the values of S and P. For example, an odds ratio lower margin of 0.5 (i.e., $r = 2$) is too low if $S = 0.75$ and $P = 0.61$, because rejection of the null hypothesis based on the odds ratio would imply $T > 0.6$, and such a conclusion is meaningless, as $P = 0.61$. Although the discussion in Section 2.4 of Chapter 2 deals with the NI margin for the mean difference, it strongly supports the use of the NI margin, which depends on the effect size of the active control (relative to placebo in the appropriate scale), in the NI trial with binary outcomes as well. As noted in Section 2.2 of Chapter 2, if there is no historical data for estimating the effect size, directly or indirectly, the active control should not be used as a control in the NI trial.

In Section 4.3, the odds is defined in such a way that a larger odds corresponds to a better outcome. However, if the odds is defined as

$$O^*(X) = (1 - X)/X$$

then a smaller odds corresponds to a better outcome. In that case, we would test the null hypothesis that

$$O^*(T)/O^*(S) \geq r$$

against the alternative hypothesis that

$$O^*(T)/O^*(S) < r$$

where the margin r (≥ 1) may be defined similarly as given by $r = [O^*(P)/O^*(S)]^\varepsilon$. These hypotheses in terms of O^* can be shown easily to be the same hypotheses in terms of O in Section 4.3, by expressing the hypotheses in terms of T and S. On the other hand, the hypotheses in terms of O* look similar to the hypotheses based on hazard ratio discussed in Section 3.4 of Chapter 3.

As shown in Section 4.2, an NI margin of 0.1 for testing the NI hypothesis based on the difference of two proportions could be questionable when S is close to 1. Similarly, an NI margin of 0.1 could be questionable when S is close to 0. This issue may be resolved by testing the NI hypothesis based on

1. The ratio of two proportions (i.e., success rates) when S is close to 0

2. The ratio of two failure rates when S is close to 1

In any case, such an NI hypothesis is almost the same as the NI hypothesis based on the odds ratio, because the odds ratio is approximately equal to the ratios when S is close to 0 (with O definition) or to 1 (with O* definition). Therefore, testing the NI hypothesis based on the odds ratio is recommended, although the odds ratio is conceptually more difficult to understand than is the difference in proportions.

As an example, in the evaluation of a diagnostic test kit for human immunodeficiency virus (HIV) compared with an approved marketed test kit based on over 15,000 samples, the estimates of the specificity for both the test (T) and comparator (S) are close to 1 (0.9980 vs. 0.9994). It appears that we could conclude NI based on either the difference or the ratio. However, the estimate of the false-positive rate for the test is more than three times that for the comparator (0.0020 vs. 0.0006). Therefore, the NI hypothesis should be formulated based on the ratio of the false-positive rates. From a practical point of view, maintaining a certain level of specificity may be tested directly without comparing with the control.

Testing the NI hypothesis based on the ratio or odds ratio would require a huge sample size when S is close to 0 or 1. In practice, the NI margin is often loosened so that the study can be conducted with a manageable sample size on the basis of showing efficacy as compared to putative placebo. In such situations, NI claims should not be made when the null hypothesis is rejected; therefore, NI could not be shown statistically from a practical point of view.

References

Garrett AD (2003). Therapeutic equivalence: fallacies and falsification. *Statistics in Medicine*, **22**:741–762.

Hung H-MJ, Wang S-J, and O'Neill RT. (2005). A Regulatory Perspective on Choice of Margin and Statistical Inference Issue in Non-Inferiority Trials. *Biometrical Journal*, **47**:28–36.

McCullagh P, and Nelder JA (1990). *Generalized Linear Models*. New York: Chapman and Hall, 31.

Ng T-H (1993). A Specification of Treatment Difference in the Design of Clinical Trials with Active Controls. *Drug Information Journal*, **27**:705–719.

Ng T-H (2001). Choice of Delta in Equivalence Testing. *Drug Information Journal*, **35**:1517–1527.

Rothmann M, Li N, Chen G, Chi GY-H, Temple R, and Tsou H-H. (2003). Design and Analysis of Non-Inferiority Mortality Trials in Oncology. *Statistics in Medicine*, **22**:239–264.

Senn S (2000). Consensus and Controversy in Pharmaceutical Statistics (with Discussion). *Journal of the Royal Statistical Society*, Series D, **49**:135–176.

Simon R (1999). Bayesian Design and Analysis of Active Control Clinical Trials. *Biometrics*, **55**:484–487.

Tsong Y, Wang S-J, Hung H-MJ, and Cui L (2003). Statistical Issues on Objectives, Designs and Analysis of Non-inferiority Test Active-Controlled Clinical Trials. *Journal of Biopharmaceutical Statistics*, **13**:29–41.

Tu D (1998). On the Use of the Ratio or the Odds Ratio of Cure Rates in Therapeutic Equivalence Clinical Trials with Binary Endpoints. *Journal of Biopharmaceutical Statistics*, **8**:263–282.

Wellek S (2005). Statistical Methods for the Analysis of Two-Arm Non-inferiority Trials with Binary Outcomes. *Biometrical Journal*, **47**:48–61.

5

Two Statistical Approaches for Testing the Noninferiority Hypothesis

5.1 Introduction

Although the discussion in this chapter is in the context of the mean difference for a continuous endpoint introduced in Chapter 2, the concept can be adapted easily to the mean ratio for a continuous endpoint and a hazard ratio for a survival endpoint, as discussed in Chapter 3, as well as the difference in proportions, the relative risk, and the odds ratio for a binary endpoint discussed in Chapter 4.

Incorporating the noninferiority (NI) margin given by Equation 2.2 in Section 2.5.2 of Chapter 2 into the hypotheses of Equations 1.3a and 1.3b in Section 1.7 of Chapter 1, we are testing

$$H_0: \ T - S \leq -\varepsilon\gamma(S - P)_h \tag{5.1a}$$

versus

$$H_1: \ T - S > -\varepsilon\gamma(S - P)_h \tag{5.1b}$$

where $\varepsilon > 0$, $\gamma \leq 1$, and $(S - P)_h$ is the effect size in the historical trials (see Section 2.5.1 of Chapter 2). As noted in Section 2.4 of Chapter 2, the choice of ε depends on the study objective and is related to percent preservation or retention. For example, for 60% preservation, $\varepsilon = 0.4$. Gamma (γ) is the discount factor discussed in Section 2.5.2 of Chapter 2. For simplicity, assume that there is only one prior study of the active control compared to a placebo. Chapter 7 will discuss situations where there are multiple studies.

In general, there are two approaches in testing the NI hypothesis given by Equation 5.1a. The first approach is known as the fixed-margin method (see Section 5.3) because the NI margin is considered a fixed constant, even though it depends on the effect size estimated from the historical data (e.g., using the lower confidence limit [LCL] for the effect size). Therefore, it is conditioned on the historical data. The second approach is known as the synthesis method (see Section 5.4) because it combines or synthesizes the data from the historical trials and the current NI trial (U.S. FDA 2010). In this approach, the effect size from the historical trials is considered as a parameter, and the test statistic takes into account the variability of estimating this parameter from historical data. Therefore, it is unconditioned on the historical data.

These two approaches have been discussed widely in the literature. See, for example, Hauck and Anderson (1999), Ng (2001), Tsong et al. (2003), Wang and Hung (2003).

5.2 One-Sided versus Two-Sided

The conventional null hypothesis of equality is typically tested at the two-sided significance level α of 0.05. A statistically significant difference only in the "right" direction indicates the superiority of the test treatment against the control treatment, which could be a placebo or an active control. Therefore, the significant level of interest is effectively $\alpha/2$, or 0.025.

The NI hypothesis is one-sided in nature. But, to be consistent, in Sections 5.3 and 5.4, the NI hypothesis will be tested at $\alpha/2$ significance level, or 0.025.

5.3 Fixed-Margin Method

Using the fixed-margin method, we replace $(S - P)_h$ with the lower limit of the two-sided $(1 - \alpha^*)100\%$ (e.g., 95%) confidence interval [or the one-sided $(1 - \alpha^*/2)100\%$ LCL] for $(S - P)_h$; that is,

$$(\widehat{S-P})_h - z_{1-\alpha^*/2} \, SD(\widehat{S-P})_h$$

where $z_{1-\alpha^*/2}$ denotes the $(1 - \alpha^*/2)100$ percentile of the standard normal distribution, and $SD(\widehat{S-P})_h$ denotes the standard deviation of the estimator $(\widehat{S-P})_h$. Therefore, the NI margin is given by

$$\delta = \varepsilon\gamma[(\widehat{S-P})_h - z_{1-\alpha^*/2} \, SD(\widehat{S-P})_h]$$

At a significance level of $\alpha/2$ (see Section 5.2), we reject H_0 if the lower limit of the $(1 - \alpha)100\%$ confidence interval [or the one-sided $(1 - \alpha/2)100\%$ LCL] for $(T - S)$ exceeds $-\delta$; that is,

$$(\widehat{T-S}) - z_{1-\alpha/2} \, SD(\widehat{T-S}) > -\delta = -\varepsilon\gamma\,[(\widehat{S-P})_h - z_{1-\alpha^*/2} \, SD(\widehat{S-P})_h] \quad (5.2)$$

or equivalently,

$$(\widehat{T-S})+\varepsilon\gamma\,(\widehat{S-P})_h > z_{1-\alpha/2}\,SD(\widehat{T-S})+\varepsilon\gamma z_{1-\alpha^*/2}\,SD(\widehat{S-P})_h \qquad (5.3)$$

where $z_{1-\alpha/2}$ denotes the $(1-\alpha/2)100$ percentile of the standard normal distribution, and $SD\,(\widehat{T-S})$ denotes the standard deviation of the estimator $(\widehat{T-S})$. If $\alpha^* = \alpha = 0.05$—that is, the 95% confidence intervals are used—then we reject H_0 if

$$\frac{(\widehat{T-S})+\varepsilon\gamma(\widehat{S-P})_h}{SD(\widehat{T-S})+\varepsilon\gamma\,SD(\widehat{S-P})_h} > z_{1-\alpha/2} = 1.96$$

The fixed-margin method is also referred to as the two confidence-interval method, since two confidence intervals are used: one to estimate the effect size in the determination of the NI margin and the other one to test the null hypothesis, or more specifically, the 95%-95% method (Hung, Wang, and O'Neil 2007). Since the lower bounds of the 95% confidence intervals are used, the effective significance levels are 97.5%. For this reason, such a fixed-margin method may be referred to as 97.5%-97.5% method.

5.4 Synthesis Method

Using the synthesis method, rewrite the hypotheses in Equations 5.1a and 5.1b, respectively, as

$$H_0: T-S+\varepsilon\gamma(S-P)_h \leq 0 \qquad (5.4a)$$

and

$$H_1: T-S+\varepsilon\gamma(S-P)_h > 0 \qquad (5.4b)$$

At a significance level of $\alpha/2$ (see Section 5.2), we reject H_0 if the lower limit of the $(1-\alpha)100\%$ confidence interval [or the one-sided $(1-\alpha/2)100\%$ LCL] for $(T-S) + \varepsilon\gamma(S-P)_h$ is greater than 0. That is,

$$(\widehat{T-S})+\varepsilon\gamma(\widehat{S-P})_h - z_{1-\alpha/2}\sqrt{Var(\widehat{T-S})+\varepsilon^2\gamma^2 Var(\widehat{S-P})_h} > 0 \qquad (5.5)$$

or equivalently,

$$\frac{(\widehat{T-S})+\varepsilon\gamma(\widehat{S-P})_h}{\sqrt{Var(\widehat{T-S})+\varepsilon^2\gamma^2 Var(\widehat{S-P})_h}} > z_{1-\alpha/2} = 1.96$$

if $\alpha = 0.05$, where "Var" stands for the variance.

5.5 Fixed-Margin Method versus Synthesis Method

If the lower limit of the 95% confidence interval (or the one-sided 97.5% LCL) for the historical effect size is used to determine the NI margin, then the denominator of the test statistic using the fixed-margin approach, but viewing the historical data unconditionally (see Section 5.3), is always larger than that of the test statistic using the synthesis method (see Section 5.4) because

$$SD(\widehat{T-S}) + \varepsilon\gamma SD(\widehat{S-P})_h \geq \sqrt{Var(\widehat{T-S}) + \varepsilon^2\gamma^2 Var(\widehat{S-P})_h}$$

Therefore, it is easier to reject the null hypothesis given by Equation 5.1a using the synthesis method than to use the fixed-margin approach if the two-sided confidence level for estimating $(S - P)_h$ is at least 95%. However, that is not necessarily true if the confidence level for estimating $(S - P)_h$ is less than 95%. For example, at the extreme, if the point estimate of $(S - P)_h$ is used (corresponding to a 0% confidence interval, i.e., $\alpha^* = 1$ and $z_{1-\alpha^*/2} = 0$), then it is easier to reject the null hypothesis given by Equation 5.1a using the fixed-margin approach than to use the synthesis method because the second term in the right side of Equation 5.3 is zero and

$$SD(\widehat{T-S}) \leq \sqrt{Var(\widehat{T-S}) + \varepsilon^2\gamma^2 Var(\widehat{S-P})_h}$$

Hauck and Anderson (1999) derived the confidence level $(1 - \alpha^*)$ to estimate $(S - P)_h$ so that the two approaches are equivalent, assuming that the constancy assumption holds (i.e., $\gamma = 1$) with no preservation (i.e., $\varepsilon = 1$). In general, using Equations 5.3 and 5.5, the two approaches are equivalent if and only if

$$z_{1-\alpha/2}\, SD(\widehat{T-S}) + \varepsilon\gamma z_{1-\alpha^*/2}\, SD(\widehat{S-P})_h = z_{1-\alpha/2}\sqrt{Var(\widehat{T-S}) + \varepsilon^2\gamma^2 Var(\widehat{S-P})_h}$$

or equivalently,   (5.6)

$$z_{1-\alpha^*/2} = z_{1-\alpha/2}\sqrt{[R/(\varepsilon\gamma)]^2 + 1} - z_{1-\alpha/2}\,R/(\varepsilon\gamma)$$

where

$$R = SD(\widehat{T-S})/SD(\widehat{S-P})_h$$

If $R = \varepsilon = \gamma = 1$ and $\alpha = 0.05$, then $\alpha^* = 0.4169$, which corresponds to 58.31% confidence level. Note that such a confidence level depends on the sample sizes (in addition to γ and ε) through the ratio of the standard deviations of estimating $(S - T)$ and $(S - P)_h$.

Incorporating Equation 5.6, one can determine the NI margin δ in Equation 5.2 instead of the confidence level $(1 - \alpha^*)$, such that the two approaches are equivalent, as follows:

$$
\begin{aligned}
\delta &= \varepsilon\gamma[(\widehat{S-P})_h - z_{1-\alpha^*/2} \, SD(\widehat{S-P})_h] \\
&= \varepsilon\gamma(\widehat{S-P})_h - \varepsilon\gamma z_{1-\alpha^*/2} \, SD(\widehat{S-P})_h \\
&= \varepsilon\gamma(\widehat{S-P})_h + z_{1-\alpha/2} \, SD(\widehat{T-S}) - z_{1-\alpha/2}\sqrt{Var(\widehat{T-S}) + \varepsilon^2\gamma^2 Var(\widehat{S-P})_h}
\end{aligned}
$$

A similar expression is also given by Hung et al. (2003).

The two approaches are intrinsically different. The fixed-margin method is conditioned on the historical data through the determination of the NI margin and controls the conditional Type I error rate in the sense of falsely rejecting the null hypothesis with the given NI margin when the NI trial is repeated (Hung, Wang, and O'Neil 2007). The synthesis method considers $(S - P)_h$ as a parameter and factors the variability into the test statistic. Thus, it is unconditional and controls the unconditional Type I error rate, in the sense of falsely rejecting the null hypothesis when the historical trials and the NI trial are repeated (Lawrence 2005).

Apart from the differences in the Type I error rate, the synthesis approach has other limitations compared to the fixed-margin method. The synthesis approach may not provide independent evidence of treatment effect from multiple NI trials to provide replication, since it uses the same historical data unconditionally (Soon et al. 2013). Furthermore, in the absence of a prespecified NI margin, it might be difficult to appropriately plan and design the NI trials.

It is understood that the Type I error rate is (1) conditional if the fixed-margin method is used and (2) unconditional if the synthesis method is used.

The equality of two means may be tested based on (1) the 95% confidence interval for the mean difference or (2) the 95% confidence intervals for the individual means. More specifically, the null hypothesis of equality of two means is rejected if (1) the confidence interval for the mean difference excludes 0 or (2) the confidence intervals for the individual means are disjointed. Schenker and Gentleman (2001) refer to the first method as the "standard" method and the second method as the "overlap" method. They show that rejection of the null hypothesis by the overlap method implies rejection by the standard method, but not vice versa. In other words, it is easier to reject the null hypothesis by the standard method than the overlap method. Although the overlap method is simple and especially convenient when lists or graphs of confidence intervals have been presented, the authors conclude that it should not be used for formal significance testing.

The standard method corresponds to the synthesis method, where one confidence interval is used, while the overlap method corresponds to the fixed-margin method, where two confidence intervals are used. From this point of view, the synthesis method rather than the fixed-margin method should be used. However, we design the NI trial conditioned on the availability of historical trials; therefore, controlling the conditional Type I error rate makes more sense than controlling the unconditional Type I error rate. Therefore, from a practical point of view, the fixed-margin method rather than the synthesis method should be used.

5.6 Efficacy

How can we conclude efficacy of the experimental treatment without a concurrent placebo? We can do so because we assume that the effect size of the active control is positive and can be estimated from the historical data.

The driving force behind the use of previous placebo-controlled studies of the active-control drug to infer efficacy of the test drug in an active-control equivalence study is the Code of Federal Regulations (CFR 1985), which states the following:

> The analysis of the study should explain why the drugs should be considered effective in the study, for example, by reference to results in previous placebo-controlled studies of the active control drug.

This statement remained unchanged as of April 1, 2013 (CFR 2013). However, it did not give any direction on how to use the previous studies to infer efficacy of the test treatment. Fleming (1987) gave a more explicit direction in the following:

> Using information on the relationship of the new drug to the active control and of the active control to no treatment, one can estimate the relationship of the new drug to no treatment and thereby obtain the desired quantitative assessment of the new drug effect.

Ng (1993) translated the last statement into formulas and proposed a test statistic for inferring the efficacy of the test drug as compared to placebo. It is simple and straightforward, and there is no need to specify δ. This is known as the synthesis method, as discussed in Section 5.4. Note that the validity of this method depends on the constancy assumption (Section 2.5.1).

Hauck and Anderson (1999) discussed the two approaches to establish the efficacy ($\varepsilon = 1$) of a test drug as compared to placebo, assuming the constant assumption ($\gamma = 1$). Indirect comparisons with placebo were also discussed by many authors (e.g., Hassalblad and Kong 2001; Julious and Wang 2008; Julious 2011; Snapinn and Jiang 2011).

5.7 Preservation versus Discounting

If we decide on a 50% preservation ($\varepsilon = 0.5$) with 20% discounting ($\gamma = 0.8$), then $\delta = 0.4(S - P)_h$. Using the fixed-margin method with the one-sided 97.5% LCL to estimate $(S - P)_h$, we reject H_0, if

$$\frac{(\widehat{T - S}) + 0.4(\widehat{S - P})_h}{SD(\widehat{T - S}) + 0.4SD(\widehat{S - P})_h} > 1.96$$

Using the synthesis method, we reject H_0 if

$$\frac{(\widehat{T - S}) + 0.4(\widehat{S - P})_h}{\sqrt{Var(\widehat{T - S}) + 0.16Var(\widehat{S - P})_h}} > 1.96$$

On the other hand, if we decide on a 60% preservation ($\varepsilon = 0.4$) with no discounting ($\gamma = 1$), then $\delta = 0.4(S - P)_h$. Therefore, different sets of preservation and discounting may result in the same NI margin leading to the same statistical test. Note that no "double credit" should be allowed—for example, concluding that the test preserves greater than 60% of the control effect with 60% discounting (or any non-zero discounting) when the null hypothesis is rejected.

Preservation dictates the size of the NI margin. The larger the percent preservation, the smaller the NI margin will be. On the other hand, discounting is used to alleviate the concern that the constancy assumption might not hold. The larger the discount, the smaller the NI margin. Therefore, preservation and discounting are two different concepts, although they are indistinguishable mathematically (Ng 2001).

5.8 Control of Type I Error Rates

It should be noted that the fixed-margin method controls the Type I error rate for testing the null hypothesis given by Equation 1.3a in Section 1.7 of Chapter 1 at the $\alpha/2$ significance level, where δ is determined by $\varepsilon\gamma$ times the one-sided $(1 - \alpha^*/2)100\%$ LCL for $(S - P)_h$. The NI margin δ is considered a fixed constant, and the Type I error rate is conditioned on the historical data. When the null hypothesis is rejected, we conclude that T is δ-no-worse-than S (see Section 1.2 of Chapter 1), that is, $T > S - \delta$. Can we conclude that T preserves greater than $(1 - \varepsilon)100\%$ of the control effect?

In principle, we would conclude that the test treatment preserves greater than $(1 - \varepsilon)100\%$ of the control effect when the null hypothesis given by

Equation 5.1a is rejected at a $\alpha/2$ significance level using either approach. However, the conditional Type I error rate using the fixed-margin method may or may not be controlled at the nominal significance level (e.g., 0.025), depending upon whether (S − P) is under- or overestimated by γ times the one-sided $(1 − \alpha^*/2)100\%$ LCL. Similarly, the unconditional Type I error rate using the synthesis method may or may not be controlled at the nominal significance level (e.g., 0.025), depending upon whether the discount factor (i.e., γ) is set appropriately. Note that the validity of the constancy assumption (see Section 5.8.1) and appropriate discount factor (see Section 5.8.2) affect the Type I error rate for both approaches.

5.8.1 With Constancy Assumption

In this subsection, we assume that the constancy assumption holds [i.e., $\gamma = 1$ or $(S − P) = (S − P)_h$]. If we decide on a preservation of 80% ($\varepsilon = 0.2$) and use the one-sided 97.5% LCL for the effect size in determining the NI margin, then $\delta = 0.2[(\widehat{S − P})_h − z_{0.975} \, SD(\widehat{S − P})_h]$. If the null hypothesis is rejected at the 0.025 level using the fixed-margin method, we then conclude that the test treatment preserves more than 80% of the effect size. However, the actual significant level will be larger than 0.025, which means the Type I error rate is inflated, if $(S − P)_h$ is overestimated. On the other hand, the actual significant level will be smaller than 0.025 if $(S − P)_h$ is underestimated.

Note that there is a 2.5% chance that $(S − P)_h$ is overestimated by the lower limit of the 95% confidence interval, and this probability increases to 5% as the confidence level decreases to 90%. When the confidence level decreases to 0%, the confidence interval shrinks to the point estimate. In that case, there is a 50% chance that $(S − P)_h$ is overestimated by the point estimate. Therefore, there is a 50% chance that the Type I error rate will be inflated if the point estimate of the effect size is used to determine δ. In general, the amount of inflation depends on how much $(S − P)_h$ is overestimated. Since the inflation of the Type I error rate is of concern, we should keep the chance of inflation low by using a small α^* (e.g., 0.05). Therefore, using the point estimate is unacceptable. Violation of the constancy assumption may further affect the Type I error rate (see Section 5.8.2).

On the other hand, the synthesis method does not depend on the choice of α^* and controls the unconditional Type I error rate. However, the unconditional Type I error rate may be inflated if the constancy assumption is violated—for example, $\gamma = 1$ is used when, in fact, $\gamma = 0.9$ [i.e., $(S − P) = 0.9(S − P)_h$].

5.8.2 Without Constancy Assumption

If we decide on a preservation of 60% ($\varepsilon = 0.4$) with 10% discounting ($\gamma = 0.9$), and we used the one-sided 97.5% LCL for the effect size in determining the NI margin, then $\delta = 0.4 \cdot 0.9 \cdot LCL$, where $LCL = [(\widehat{S - P})_h − z_{0.975} \, SD(\widehat{S - P})_h]$.

If the null hypothesis is rejected at the 0.025 level, we then conclude that the test treatment preserves more than 60% of the effect size. However, the actual significant level will be larger (or smaller) than 0.025 if the $(S - P)$ is overestimated (or underestimated) by 0.9 times LCL. Note that two factors affect the actual nominal significance level. The first factor deals with the discounting. The second factor involves estimating the effect size of the standard therapy in the historical trials [i.e., $(S - P)_h$; not the current NI trial] by LCL.

If the discounting factor of 0.9 is correct, that is, $(S - P) = 0.9(S - P)_h$, then the discussion in Section 5.8.1 regarding the fixed-margin approach is applicable to the second factor; however, if the discounting factor of 0.9 is incorrect, that is, $(S - P) \neq 0.9(S - P)_h$, then the Type I error rate will be affected in different ways. For example, if $(S - P) = 0.95(S - P)_h$ (the discounting of 10% is too high), then there will be no inflation of the Type I error rate, even though $(S - P)_h$ may be slightly overestimated by LCL, as long as $(S - P)$ is not overestimated by 0.9 times LCL. In other words, there will be no inflation of the Type I error rate provided an overestimation of $(S - P)_h$ by LCL is completely offset by overdiscounting.

With the synthesis method, if $(S - P) = 0.9(S - P)_h$ and 10% discounting is used, then there is no inflation of the Type I error rate; however, if $(S - P) < 0.9(S - P)_h$ (a 10% discount is too small), the Type I error rate will be inflated.

References

Code of Federal Regulations, 21 CFR 314.126 (1985). Adequate and Well-Controlled Studies.

Code of Federal Regulations, 21 CFR 314.126 (2013). Adequate and Well-Controlled Studies http://www.accessdata.fda.gov/scripts/cdrh/cfdocs/cfcfr/CFRSearch.cfm?fr=314.126 (Accessed: March 3, 2014).

Fleming TR (1987). Treatment Evaluation in Active Control Studies. *Cancer Treatment Reports*, 71:1061–1065.

Hassalblad V and Kong DF (2001). Statistical methods for comparison to placebo in active-control trials. *Drug Information Journal*, 35:435–449.

Hauck WW and Anderson S (1999). Some Issues in the Design and Analysis of Equivalence Trials. *Drug Information Journal*, 33:109–118.

Hung HMJ, Wang SJ, and O'Neil RT (2007). Issues with Statistical Risks for Testing Methods in Noninferiority Trials Without a Placebo Arm. *Journal of Biopharmaceutical Statistics*, 17:201–213.

Hung HMJ, Wang SJ, Tsong Y, Lawrence J, and O'Neill RT (2003). Some Fundamental Issues with Noninferiority Testing in Active Controlled Clinical Trials. *Statistics in Medicine*, 22:213–225.

Julious SA (2011). The ABC of Non-inferiority Margin Setting from Indirect Comparisons. *Pharmaceutical Statistics*, 10:448–453.

Julious SA and Wang SJ (2008). Issues with Indirect Comparisons In Clinical Trials Particularly with Respect to Non-inferiority Trials. *Drug Information Journal,* **42**(6):625–633.

Lawrence J (2005). Some Remarks about the Analysis of Active-Control Studies. *Biometrical Journal,* **47**(5):616–622.

Ng T-H (1993). A Specification of Treatment Difference in the Design of Clinical Trials with Active Controls, *Drug Information Journal,* **27**:705–719.

Ng T-H (2001). Choice of Delta in Equivalence Testing. *Drug Information Journal,* **35**:1517–1527.

Schenker N and Gentleman JF (2001). On Judging the Significance of Differences by Examining the Overlap Between Confidence Intervals. *The American Statistician,* **55**(3):182–186.

Snapinn S and Jiang Q (2011). Indirect Comparisons in the Comparative Efficacy and Non-inferiority Settings. *Pharmaceutical Statistics,* **10**:420–426.

Soon G, Zhang Z, Tsong Y, and Nie L (2013). Assessing Overall Evidence from Noninferiority Trials with Shared Historical Data. *Statistics in Medicine,* **32**(14):2349–2363.

Tsong Y, Wang S-J, Hung H-MJ and Cui L (2003). Statistical issues on objectives, designs and analysis of non-inferiority test active controlled clinical trials. *Journal of Biopharmaceutical Statistics,* **13**:29-41.

U.S. Food and Drug Administration (2010). Draft Guidance for Industry: Non-inferiority Clinical Trials www.fda.gov/downloads/Drugs/GuidanceCompliance RegulatoryInformation/Guidances/UCM202140.pdf (Accessed: August 25, 2013).

Wang S-J and Hung H-MJ (2003). Assessment of Treatment Efficacy in Non-inferiority Trials. *Controlled Clinical Trials,* **24**:147–155.

6

Switching between Superiority and Noninferiority

6.1 Introduction

Switching between superiority and noninferiority (NI) is attractive in active equivalence control studies. It reduces the simultaneous testing of both hypotheses using a one-sided confidence interval. There was considerable interest in this topic in the late 1990s by the regulatory authority in Europe (e.g., EMEA/CPMP 2000), as well as among pharmaceutical statisticians (e.g., Phillips et al. 2000). Dunnett and Gent (1996) and Morikawa and Yoshida (1995) showed that multiplicity adjustment is not necessary by using the intersection-union (IU) and the closed-testing (CT) principles, respectively.

Although there is no inflation of the Type I error rate, Ng (2003) cautioned against such simultaneous testing for NI and superiority in a confirmatory evaluation. Simultaneous testing is exploratory in nature because it may be considered as testing only one null hypothesis that depends on the outcome of the trial, such as the lower confidence limit. Therefore, a finding of superiority in simultaneous testing is less credible than a finding of superiority in a superiority trial where no switching from superiority to NI is allowed. Depending upon how the U.S. Food and Drug Administration's (FDA's) general requirement of at least two adequate and well-controlled clinical trials is interpreted (see Section 6.2.5), a finding of superiority in simultaneous testing may or may not be used as one of the two trials required to claim superiority (see Section 6.3.2).

In addition, there are other concerns. Simultaneous testing of both hypotheses allows a test treatment that is expected to have the same effect as an active control to claim superiority by chance alone, without losing the chance of showing NI. This would lead to a higher number of erroneous claims of superiority, compared with the situation where only one null hypothesis is to be tested, because of the following. If only one null hypothesis is to be tested, if researchers expect the test treatment to have the same effect as an active control, they will likely choose to test NI rather than superiority. However, with simultaneous testing, superiority will be tested, regardless of the expectation. Therefore, more test treatments that are expected to have the same effect as an active control would be tested for superiority with simultaneous testing than would be if only one null hypothesis were to be tested. Consequently, simultaneous testing will lead to (1) more erroneous claims of superiority, although the Type I error rate remains the same; and (2) a higher false discovery rate because the prior probability is increased (Ng 2007), which is one of the reasons why most published research findings are false (Ioannidis 2005). The details will be discussed in Section 6.3.3.

Section 6.2 presents background information, including (1) the Committee for Proprietary Medicinal Products (CPMP) points-to-consider document on switching between superiority and NI (see Section 6.2.1) and (2) simultaneous tests for NI and superiority (see Section 6.2.2). Section 6.3 presents the statistical issues with simultaneous testing, including increases in the false discovery rate. Decision-theoretic views are presented in Section 6.4, followed by controlling the Type I error rate of superiority claims conditioned on establishing the NI in Section 6.5. Discussions and concluding remarks are given in Section 6.6.

It should be noted that in this chapter, we (1) use the fixed-margin method discussed in Section 5.3 in Chapter 5, and (2) assume that the same analysis set is used in the NI testing and the superiority testing for simplicity, although it is most likely not the case in practice, as discussed in Chapter 10.

6.2 Background

6.2.1 Switching between Superiority and Noninferiority

In 2000, the European Agency for the Evaluation of Medicinal Products, Committee for Proprietary Medicinal Products (EMEA/CPMP 2000) issued a document titled "Points to Consider on Switching between Superiority and Noninferiority."

Basically, in an NI trial, if the null hypothesis is rejected, we can proceed to test it for superiority. There is no multiplicity issue because the test procedure is closed. In a superiority trial, if we fail to reject the null hypothesis, we can proceed to test it for NI.

The document asserts that there is no multiplicity issue. However, it does point out the issue of post hoc specification of δ, if it is not already specified.

6.2.2 Simultaneous Tests for Noninferiority and Superiority

Switching the objective between superiority and NI means simultaneously testing using a one-sided $(1 - \alpha)$ 100% lower confidence interval (CI), as shown in Figure 6.1. The axis is the mean difference of the test treatment minus that of the standard therapy. If the lower limit of the CI is greater

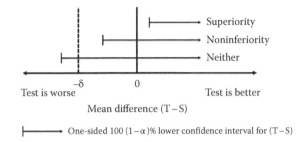

FIGURE 6.1
Simultaneous testing for noninferiority and superiority.

than 0, then superiority is shown. If it is between $-\delta$ and 0, then NI is shown. Otherwise, neither NI nor superiority is shown.

Such simultaneous testing is discussed by Dunnett and Gent (1996) and Morikawa and Yoshida (1995). Both papers argue that a multiplicity adjustment is not necessary. The first paper uses the IU principle; that is, we conclude superiority if both hypotheses for NI and superiority are rejected. The second paper uses the CT principle, in which we test the null hypothesis for superiority when the intersection of hypotheses for NI and superiority is tested and rejected.

In October 1998, the Statisticians in the Pharmaceutical Industry (PSI) (Phillips et al. 2000) organized a discussion forum in London. One of the questions posed was "Is simultaneous testing of equivalence and superiority acceptable?" There was no consensus in the discussion. Some participants felt that in a superiority trial, if we fail to reject the null hypothesis of equality, no claim of equivalence/NI could be made. Others felt that it is permissible to proceed to test for NI, and they cited the second paper (i.e., Morikawa and Yoshida 1995).

6.2.3 Assumptions and Notation

We assume that the response variable follows a normal distribution with a common variance σ^2 and that a larger response variable corresponds to a better treatment. For a given d, define the null hypothesis $H_0(d)$ as the test treatment being worse than the standard therapy by d or more; that is,

$$H_0(d): T \leq S - d$$

and the alternative hypothesis $H_1(d)$ is the complement of the null hypothesis; that is,

$$H_1(d): T > S - d$$

To test for NI, set $d = \delta$; to test for superiority, set $d = 0$. These hypotheses are shown graphically in Figures 6.2a and 6.2b, respectively, where the axis

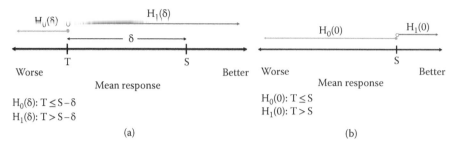

FIGURE 6.2
(a) Test for noninferiority ($d = \delta$). (b) Test for superiority ($d = 0$).

represents the mean response. Finally, we assume that all null hypotheses are tested at the same α level.

6.2.4 Hypothesis Testing versus Interval Estimation

It is well known that we can test the null hypothesis $H_0(d)$ at significance level α by constructing a one-sided $(1 - \alpha)100\%$ lower CI for $T - S$, and rejecting the null hypothesis if and only if the CI excludes $-d$, as shown in Figure 6.3. In this figure, the axis represents the mean difference of the test treatment minus the standard therapy.

If the lower limit of the CI is L, then $H_0(d)$ will be rejected, for all $-d < L$. However, we should be cautious not to fall into the trap of post hoc specification of the null hypothesis in the sense of specifying "d" as the lower limit of the 95% CI after seeing the data. For example, in a premarket notification 510(k) submission to the U.S. FDA in December 2000, it was stated that when using a one-sided 95% CI, the true mean of this population is greater than 83.6%. Presumably, 83.6% is the one-sided 95% lower confidence limit. Since 83.6% is not prespecified, the conclusion that the true mean of this population is greater than 83.6% cannot be treated as confirmatory in the sense of hypothesis testing; otherwise, it would be just like testing the null hypothesis that can just be rejected (Pennello, Maurer, and Ng 2003). Such post hoc

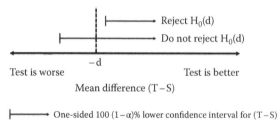

FIGURE 6.3
Use of confidence interval in hypothesis testing.

specification of the null hypothesis is exploratory and is unacceptable for confirmatory testing for the following reason. If the study is repeated with the same study design, sample size, etc., then the unconditional probability of rejecting the same null hypothesis again (in the sense of considering the threshold value being random) is only 50%.

As we know, in hypothesis testing, when the null hypothesis $H_0(d)$ is rejected, we conclude that $T - S > -d$. We note that it is most likely that $T - S$ is considerably larger than $-d$ (but we never know how much larger); otherwise, the study would not have had enough power to reject the null hypothesis.

On the other hand, if the $100(1 - \alpha)\%$ lower confidence limit is L, we want to refrain from concluding that $T - S > L$ because doing so is similar to post hoc specification of the null hypothesis.

6.2.5 Two Adequate and Well-Controlled Clinical Trials

The U.S. FDA's general requirement of at least two adequate and well-controlled clinical trials may be interpreted in two different ways:

1. Two independent confirmatory trials are conducted more or less in parallel.
2. Two trials are conducted sequentially. The results of the first trial may be used to design the second trial.

These two different interpretations are discussed by Maurer (Pennello, Maurer, and Ng 2003), respectively, as (1) two "identical" trials run in parallel and (2) two trials are conducted sequentially.

6.3 Statistical Issues with Simultaneous Testing

6.3.1 Type I Error Control and Logical Flaws

Although simultaneous testing for NI and superiority is accepted by the EMEA/CPMP (2000) and the U.S. FDA (2010), this can lead to the acceptance of testing several nested null hypotheses simultaneously, which is not desirable. Two such arguments against simultaneous testing are presented in this subsection.

Instead of simultaneous testing for NI and superiority, simultaneous testing of two null hypotheses, $H_0(d_1)$ and $H_0(d_2)$, for any $d_1 > d_2$ may be performed as shown in Figure 6.4a. If the lower limit exceeds $-d_2$, we reject $H_0(d_2)$. If the lower limit is between $-d_1$ and $-d_2$, we reject $H_0(d_1)$, but not $H_0(d_2)$. Otherwise, we reject neither of the null hypotheses. According

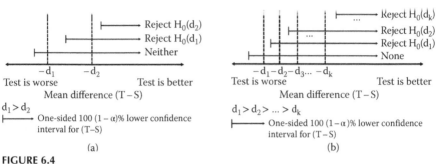

FIGURE 6.4

(a) Simultaneous testing of $H_0(d_1)$ and $H_0(d_2)$. (b) Simultaneous testing of $H_0(d_1)$, $H_0(d_2)$, ..., $H_0(d_K)$.

to Dunnett and Gent (1996) and Morikawa and Yoshida (1995), with this approach, no multiplicity adjustment is necessary. If we can simultaneously test two hypotheses without adjustment, there is no reason why we can't simultaneously test three nested null hypotheses. In fact, we can simultaneously test any number, say k, of nested hypotheses, $H_0(d_1)$, ..., $H_0(d_k)$, without adjustment, as shown in Figure 6.4b for $d_1 > d_2 > ... > d_k$. Operationally, such simultaneous testing of k nested hypotheses is like specifying one null hypothesis to be tested after seeing the data. For example, if the one-sided $100(1 - \alpha)\%$ lower confidence limit falls between $-d_j$ and $-d_{j+1}$ for some j, then we test $H_0(d_j)$. Since we can simultaneously test as many nested hypotheses as we like without adjustment, we can choose k large enough and $d_1 > d_2 > ... > d_k$ so that (1) the one-sided $100(1 - \alpha)\%$ lower confidence limit exceeds $-d_1$ almost with certainty, and (2) the difference between the two adjacent d's is as small as we like. Therefore, simultaneous testing of these k nested hypotheses is similar to post hoc specification of the null hypothesis that can just be rejected.

To reiterate the problem, if we accept simultaneous testing for NI and superiority without some kind of adjustment (but not a multiplicity adjustment for the Type I error rate), then we have no reason not to accept simultaneous testing of three, four, or any number of k nested null hypotheses, which contradicts the fact that post hoc specification of the null hypothesis (in the sense discussed in Section 6.2.4) is unacceptable. Therefore, accepting simultaneous testing of NI and superiority on the basis of no inflation of the Type I error rate is, logically, a flaw.

There is nothing wrong with the IU and CT principles. In fact, the probability of rejecting at least one true null hypothesis is controlled at α when many nested null hypotheses are tested simultaneously. In other words, there is no inflation of the Type I error rate. This can be shown easily as follows (see also Hsu and Berger 1999). If $T - S$ is in $(-d_j, -d_{j+1})$ for some j, then $H_0(d_i)$ is false for $i \leq j$ and true for all $i > j$. It follows that

$$\text{Pr[Rejecting at least one true null hypothesis]}$$
$$\leq \text{Pr[Rejecting } H_0(d_{j+1}) \,|\, T - S = -d_{j+1}] = \alpha$$

where "Pr" stands for probability. If we accept simultaneous testing of NI and superiority because the Type I error rate is controlled, why don't we accept simultaneous testing of many nested null hypotheses as well? Here is the problem with simultaneous testing of many nested null hypotheses. If we simultaneously test many nested null hypotheses, we will have a low probability of confirming the findings of such testing. For example, in the first trial, if $H_0(d_j)$ is rejected but $H_0(d_{j+1})$ is not and the same trial is repeated, then the unconditional probability (in the sense of considering $H_0(d_j)$ being random) that $H_0(d_j)$ will be rejected in the second trial could be as low as 50% as the number of nested hypotheses approaches infinity (see Section 6.2.4). Therefore, accepting simultaneous testing of NI and superiority on the basis of the Type I error rate being controlled is, logically, a flaw.

6.3.2 An Assessment of the Problems

How would we assess the problems when two nested null hypotheses are tested simultaneously, in particular, when NI and superiority are tested simultaneously? One way is to assess the probability of confirming the finding from the first trial in the presumed second independent trial relative to that of testing for NI. To do so, we assume that the variance is known and let

$$\theta = T - S$$

For a fixed d, let $f_d(\theta)$ be the power function for testing the null hypothesis $H_0(d)$; that is, $f_d(\theta) = \Pr[\text{Rejecting } H_0(d)|\theta]$.

These power functions are shown graphically in Figure 6.5a, for $d = \delta\ (= 2)$ and 0, where $\alpha = 0.025$. The sample size is such that $f_\delta(0)$ is 0.8; that is, the study has 80% power to conclude NI at $\theta = 0$.

Suppose we test one null hypothesis $H_0(\delta)$ and we reject it. If the same trial is repeated independently, then the probability of rejecting the null hypothesis again in the second trial is given by the solid line in Figure 6.5a, which is denoted by $f_\delta(\theta)$.

Suppose we test $H_0(\delta)$ and $H_0(0)$ simultaneously and $H_0(\delta)$ or $H_0(0)$ is rejected. If the same trial is repeated independently, then the probability (as a function of θ, but given that $H_0(\delta)$ is rejected) of rejecting the same hypothesis again in the second trial is given by

Pr[$H_0(\delta)$ is rejected but not $H_0(0)$ | $H_0(\delta)$ is rejected] $\times$ Pr[$H_0(\delta)$ is rejected in the second trial]

plus Pr[$H_0(0)$ is rejected | $H_0(\delta)$ is rejected] $\times$ Pr[$H_0(0)$ is rejected in the second trial]

$$= \{[f_\delta(\theta) - f_0(\theta)]\,/\,f_\delta(\theta)\} \cdot f_\delta(\theta) + [f_0(\theta)\,/\,f_\delta(\theta)] \cdot f_0(\theta)$$

$$= [1 - w(\theta)] \cdot f_\delta(\theta) + w(\theta) \cdot f_0(\theta)$$

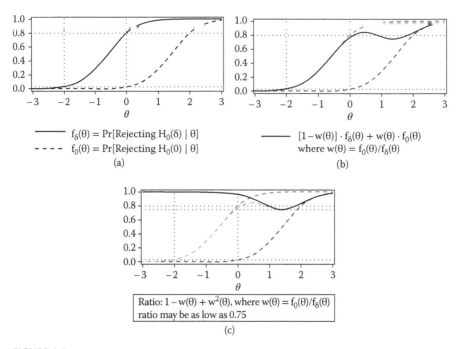

FIGURE 6.5
(a) Power functions of testing one null hypothesis (2nd trial). (b) Power function of simultaneous testing (2nd trial). (c) Relative power function of simultaneous testing (2nd trial).

where $w(\theta) = f_0(\theta)/f_\delta(\theta)$. Note that this power function is a weighted average of the two power functions and is shown graphically by the solid line in Figure 6.5b. Taking the ratio of this power function over the power function of the second trial when only $H_0(\delta)$ is tested—that is, $f_\delta(\theta)$—we have

$$1 - w(\theta) + w^2(\theta)$$

This ratio is shown graphically in Figure 6.5c. It can be shown that this ratio may be as low as 0.75. In other words, there may be a 25% reduction in power in confirming the finding of simultaneous testing for NI and superiority.

It should be noted that the main purpose of the second trial is to determine if the conclusion drawn from the first trial, where nested hypotheses are tested simultaneously, is credible, rather than to serve as one of the two adequate and well-controlled clinical trials. In other words, we are making an assessment on whether the first trial may be used as one of the confirmatory trials in the first interpretation (see Section 6.2.5). For example, one should not believe the conclusion based on testing the null hypothesis can just be rejected because there is only a 50% chance (unconditional on the outcome of the first trial) of the null hypothesis being rejected again if the trial is repeated with the same sample size. Simultaneous testing for NI and superiority is exploratory in nature, in the sense of testing one null hypothesis that is data dependent [i.e., testing $H_0(\delta)$ if the lower confidence limit < 0, and $H_0(0)$ otherwise]. Therefore, the study should not be used as one of the two independent confirmatory trials for the first interpretation to claim superiority. On the other hand, it may be used as one of the two well-controlled clinical trials required by the U.S. FDA to claim superiority in the second interpretation because it is not used as a confirmatory trial.

6.3.3 Simultaneous Testing of Noninferiority and Superiority Increases the False Discovery Rate

This subsection shows that simultaneous testing of NI and superiority (see Section 6.2.2) would increase the false discovery rate as shown by Ng (2007). Rejection of the null hypothesis in a one-sided test is called discovery by Soric (1989), who is concerned about the proportion of false discoveries in the set of declared discovery. Such a proportion corresponds to the conditional probability of a false rejection for a specific null hypothesis given that this null hypothesis is rejected. This conditional probability may be considered from a Bayesian perspective as the posterior distribution. This posterior distribution depends on the Type I error rate (α), the statistical power ($1 - \beta$), and prior distribution (i.e., the probability that the null hypothesis is true). More specifically,

$$\Pr[H_0 \text{ is true} \,|\, H_0 \text{ is rejected}] = \frac{\Pr[H_0 \text{ is true}]\,\alpha}{\Pr[H_0 \text{ is true}]\,\alpha + \Pr[H_1 \text{ is true}]\,(1-\beta)}$$

Benjamini and Hochberg (1995) define the term false discovery rate (FDR) as the expected proportion of errors among the rejected hypotheses.

To envision the concern of simultaneous testing for NI and superiority, let us consider what would happen under the following two situations:

Situation 1: Testing only one null hypothesis

Situation 2: Simultaneous testing for NI and superiority

Suppose that there are 2000 products of which 1000 have the same efficacy as the active control (call this category A), while the other 1000 products are better than the active control (call this category B). To conduct a confirmatory trial, the sponsors make a preliminary assessment of their products. Based on the preliminary assessment, the products may be grouped into category A* (same efficacy) and category B* (better). Assume that the sponsor conducts an NI trial for a product in category A* and a superiority trial for a product in category B* in situation 1 and simultaneously tests for NI and superiority, regardless of the preliminary assessment in situation 2.

TABLE 6.1

A Comparison between the Two Situations: Testing for Superiority

Category	Number of Products	Situation 1: Testing One Hypothesis		Situation 2: Testing Two Hypotheses	
		Number Tested	Number Rejected	Number Tested	Number Rejected
A: H_0 is true (T = S)	1000	200	5[a]	1000	25[a]
B: H_1 is true (T > S)	1000	800	720[b]	1000	900[b]
Total	2000	1000	725	2000	925
Proportion of true H_0		0.2		0.5	
False discovery rate			1/145		1/37

[a] Assuming Type I error rate of 0.025.
[b] Assuming 90% power for detecting superiority.

Suppose that the preliminary assessment has a 20% error rate for both categories. Then, in situation 1, 200 products in category A will be tested for superiority, and 800 products in category B will be tested for superiority (see Table 6.1). On the other hand, in situation 2, all 2000 products will be tested for superiority.

In situation 1, we expect to falsely claim superiority for 5 products from category A, as compared to 25 products in situation 2, because 1000 products will be tested for superiority in situation 2, whereas only 200 products will be tested for superiority in situation 1. Although the error rate is the same in both situations, more products in category A would be tested in situation 1 than in situation 2, resulting in more erroneous claims of superiority.

Note that 1000 products will be tested for superiority in situation 1 compared with 2000 products in situation 2. Furthermore, for those products tested for superiority, the proportion of products in category A (i.e., true null hypothesis) in situation 1 is 0.2 compared with 0.5 in situation 2. Therefore, testing two hypotheses would result in an increase in the proportion of true null hypotheses. Since the proportion of true null hypotheses is the prior distribution in the Bayesian paradigm, the FDR increases from 1/145 to 1/37, as shown in Table 6.1.

6.4 Decision-Theoretic Views

Koyama and Westfall (2005) (1) summarized the issues discussed in Section 6.3.2 and (2) compared the two strategies discussed in Section 6.3.3 using a

Bayesian decision-theoretic standpoint. These issues and the two strategies were originally discussed by Ng (2003). The two strategies are (1) testing one hypothesis based on the preliminary assessment (referred to as Ng), and (2) simultaneous testing of both hypotheses, regardless of the preliminary assessment, or "test both" (referred to as TB).

There are three components in their studies:

1. A "normal with an equivalence spike" model is used as the prior distribution of the effect size, which is indexed by three parameters: the proportion of spike equivalence (p) and the mean (m) and standard (s) deviations of the normal distribution.

2. The probability of selecting the NI test (p_s) as a function of the effect size.

3. The loss matrix indexed by one parameter (c).

Five values are used for each of the five parameters (i.e., p, m, s, p_s, c), resulting in $5^5 = 3125$ combinations of these parameter values for the Ng method and $5^4 = 625$ for the TB method, because p_s is irrelevant in the TB method. For a given set of parameters, an optimal critical value may be computed with the associated minimum expected loss. For each parameter, comparisons between the two methods are based on the loss relative to this minimum expected loss. Furthermore, the comparisons are based on the median of loss for other parameter values not in the comparisons. For example, the median of 625 values for a given p_s value for the Ng method is compared with the median of 625 values for the TB method.

The authors recommended always testing NI and superiority simultaneously in light of very high "winning" percentages of the TB method. However, such a recommendation is not warranted in a confirmatory trial because (1) changing the parameter values and/or the loss matrix could turn it around; (2) the error rates of 80%, 60%, and 50% (corresponding to the parameter values of 0.2, 0.4, and 0.5 for p_s, respectively) in the preliminary assessment are unrealistic; and (3) the comparisons are based on the median of loss for other parameter values (Ng 2007).

6.5 Conditional Type I Error Rate in Superiority Testing

To alleviate the concerns raised by Ng (2003, 2007) as discussed in Section 6.3 by switching the objective from NI to superiority, Yuan, Tong, and Ng (2011) proposed to control the conditional Type I error rate of the second-step superiority test at the nominal significance level of α. This leads to testing superiority at a significance level lower than α, thus decreasing the incidence of erroneous claims of superiority.

For simplicity, to derive the conditional Type I error rate, assuming σ_t^2 and σ_c^2 are known, let $\Phi(\bullet)$ and z_α be the cumulative distribution function (CDF) and the upper α-quantile of a standard normal distribution, respectively. Let $\sigma^2 = \sigma_t^2/n_t + \sigma_c^2/n_c$, where n_t and n_c denote the sample sizes for the test and control arms, respectively. Let $W = \sum X_i/n_t - \sum Y_j/n_c$, where X_i and Y_j are the individual values of the primary endpoint for the test treatment and control, respectively, for $i = 1,\ldots, n_t$ and $j = 1,\ldots, n_c$. Let the first-step NI hypothesis be tested at a significance level of α_1 so that the null hypothesis for NI is rejected when $W > -\delta + z_{\alpha_1}\sigma$. In addition, let the second-step superiority hypothesis be tested at a significance level of α_2 so that the null hypothesis for superiority is rejected when $W > z_{\alpha_2}\sigma$. If $\alpha_2 \leq \alpha_1$, then the conditional Type I error rate Ψ for the second-step superiority test T_2 is given by

$$\Psi = \sup_{\mu_t - \mu_c \leq 0} P(W > z_{\alpha_2}\sigma \mid W > -\delta + z_{\alpha_1}\sigma) = \alpha_2 / \Phi(-z_{\alpha_1} + \delta/\sigma) \qquad (6.1)$$

See Yuan, Tong, and Ng (2011) for the derivation of Ψ. To control the conditional Type I error rate of T_2 at the nominal significance level of α, set $\Psi = \alpha$ and solve for α_2. We have $\alpha_2 = \alpha\Phi(-z_{\alpha_1} + \delta/\sigma)$. Therefore, the testing procedure is as follows: (1) perform the first step NI test as usual with Type I error rate $\alpha_1 = \alpha$, and (2) perform the conditional superiority test with Type I error rate $\alpha_2 = \alpha\Phi(-z_\alpha+\delta/\sigma)$ instead of the nominal level of α. Noting that $\alpha_2 < \alpha$, in some sense, you pay a price for switching from NI to superiority.

In practice, the variance is unknown and has to be estimated. Although it is "natural" to use the formula given by Equation 6.1, with σ being replaced by its estimate when it is not known, it is not appropriate to do so because this formula is derived based on known variances. With equal and unknown variances, the formula $\alpha_2 = \alpha\overline{F}_{v,\delta/\sigma}(t_{v,\alpha})$ was derived by Yuan, Tong, and Ng (2011), where $\overline{F}_{v,\theta}(\cdot) = 1 - F_{v,\theta}(\cdot)$ and $F_{v,\theta}(\cdot)$ denote the CDF of the noncentral t-distribution with v degrees of freedom and noncentrality parameter θ. To be conservative, $\alpha_2 = \alpha^2$ may be used, where σ approaches infinity.

6.6 Discussion and Concluding Remarks

From a Bayesian perspective, $\Pr[H_0$ is true $\mid$ null is rejected] in situation 1 (see Section 6.3.3), with a 20% error rate in the preliminary assessment, is equal to 0.2, which increases to 0.5 in situation 2, so that the FDR in situation 1 is less than the FDR in situation 2. Therefore, testing two hypotheses would result in an increase in FDR as compared to testing one hypothesis. In general, it is straightforward to show that $\Pr[H_0$ is true $\mid$ null is rejected] in situation 1 is less than $\Pr[H_0$ is true $\mid$ null is rejected] in situation 2, so that the FDR in situation 1 is less than the FDR in situation 2, provided the sum of the error rates in the preliminary assessment is less than 1.

Ioannidis (2005) argued that most published research findings are false and discussed many factors that influence this problem. One of the factors is the ratio of the number of "true relationships" to "no relationships" among those tested in the field. This corresponds to the ratio of the number of products in category B to the number of products in category A (see Section 6.3.3) that could be "translated" into the proportion of true null hypotheses. Frequentist controls the Type I error rate, while Bayesian evaluates the FDR. Therefore, the Type I error rate should not be the sole criterion in multiple testing problems, and FDR should be taken into consideration (Ng 2007).

NI trials are conducted with the belief that the experimental treatment has the same effect as the active control. If NI and superiority are tested simultaneously, without any adjustment, there is a concern of erroneously concluding that the experimental treatment is superior over the active control. It is true that there is always a 2.5% chance of erroneously concluding the experimental treatment is superior over the active control in a superiority trial (assuming the null hypothesis is tested at the 2.5% level). However, for an experimental treatment that is expected to have the same effect as the active control, superiority testing will not be performed if no simultaneous testing is allowed (Ng 2003).

One way to alleviate (but not eliminate) the concern is to decrease the size of the test for superiority. In extreme situations (e.g., decreasing the size of the test for superiority to, say, 0.0001), the concern may be eliminated to a minimal level, but it essentially reduces to testing for NI (Ng 2003).

The downside of situation 1 is that we may fail to claim superiority for more products from category B than we would miss in situation 2. For example, assuming 90% power for detecting superiority for products from category B and 20% error rate of preliminary assessment, in situation 1, we expect to fail to claim superiority for 280 products, compared with 100 products in situation 2 (see Table 6.1). Therefore, there is a tradeoff between situations 1 and 2. However, if the preliminary assessment is fairly accurate (which it should be if we are designing a confirmatory trial), then the tradeoff will not be as large (Ng 2003).

Although erroneously claiming superiority does not result in letting ineffective products in the market, one has to be "fair" to the "active control." It is well known that post hoc specification of the null hypothesis in the context of multiple endpoints (that is, looking for the extreme result among multiple endpoints and then testing that one) would inflate the Type I error rate. On the other hand, simultaneous testing of many nested null hypotheses can be considered post hoc specification of the null hypothesis in the sense of choosing the "right" one to test. However, there is no inflation of the Type I error rate in such testing, as shown in Section 6.3.1, because the parameter space is one-dimensional as opposed to multidimensional, which is the case with multiple endpoints (Ng 2003).

Although there is no inflation of the Type I error rate, simultaneous testing of many nested null hypotheses is problematic in a confirmatory trial, because the probability of confirming the findings in a second trial would approach 0.5 as the number of nested null hypotheses approaches infinity. There is a concern with erroneous conclusions of superiority in simultaneous testing for NI and superiority. Such a concern would diminish if only one null hypothesis is tested, because NI trials rather than superiority trials would be conducted for experimental treatments that are expected to have the same effect as the active control. This is a good example of how there might be problems other than the inflation of the Type I error rate in multiple testing (Ng 2003).

In a confirmatory trial, we usually test one and only one prespecified primary null hypothesis and post hoc specification of the null hypothesis in the sense of specifying "d" after seeing the data is exploratory, and therefore unacceptable. Simultaneous testing of many nested null hypotheses is problematic, although there is no inflation of the Type I error rate. Simultaneous testing for NI and superiority may be viewed as an initial step toward exploratory analysis and thus, may be best used cautiously in confirmatory evaluation (Ng 2003).

References

Benjamini Y and Hochberg Y (1995). Controlling the False Discovery Rate: A Practical and Powerful Approach to Multiple Testing. *Journal of the Royal Statistical Society, Series B: Methodological*, 57:289–300.

Dunnett CW and Gent M (1996). An Alternative to the Use of Two-Sided Tests in Clinical Trials. *Statistics in Medicine*, 15:1729–1738.

European Agency for the Evaluation of Medicinal Products, Committee for Proprietary Medicinal Products (2000). Points to Consider on Switching Between Superiority and Noninferiority (http://www.ema.europa.eu/docs/en_GB/document_library/Scientific_guideline/2009/09/WC500003658.pdf) (Accessed: August 25, 2013).

Hsu J and Berger R (1999). Stepwise Confidence Intervals without Multiplicity Adjustment for Dose-Response and Toxicity Studies. *Journal of the American Statistical Association*, 94:468–482.

Ioannidis JPA (2005). Why Most Published Research Findings Are False. *PLoS Medicine* 2(8):e124, pp. 0690–0701.

Koyama T and Westfall PH (2005). Decision-Theoretic Views on Simultaneous Testing of Superiority and Noninferiority. *Journal of Biopharmaceutical Statistics*, 15:943–955.

Morikawa T and Yoshida M (1995). A Useful Testing Strategy in Phase III Trials: Combined Test of Superiority and Test of Equivalence. *Journal of Biopharmaceutical Statistics*, 5:297–306.

Ng T-H (2003). Issues of Simultaneous Tests for Non-inferiority and Superiority. *Journal of Biopharmaceutical Statistics*, **13**:629–639.

Ng T-H (2007). Simultaneous Testing of Noninferiority and Superiority Increases the False Discovery Rate. *Journal of Biopharmaceutical Statistics*, **17**:259–264.

Pennello G, Maurer W, and Ng T-H (2003). Comments and Rejoinder on "Issues of Simultaneous Tests for Non-inferiority and Superiority." *Journal of Biopharmaceutical Statistics*, **13**:641–662.

Phillips A, Ebbutt A, France L, and Morgan D (2000). The International Conference on Harmonization Guideline "Statistical Principles for Clinical Trials": Issues in Applying the Guideline in Practice. *Drug Information Journal*, **34**:337–348.

Soric B (1989). Statistical Discoveries and Effect Size Estimation. *Journal of the American Statistical Association*, **84**:608–610.

U.S. Food and Drug Administration (2010). Draft Guidance for Industry: Non-inferiority Clinical Trials (http://www.fda.gov/downloads/Drugs/GuidanceComplianceRegulatoryInformation/Guidances/UCM202140.pdf) (Accessed: August 25, 2013).

Yuan J, Tong T, and Ng T-H (2011). Conditional Type I Error Rate for Superiority Test Conditional on Establishment of Noninferiority in Clinical Trials. *Drug Information Journal*, **45**:331–336.

7

Multiple Historical Studies and Meta-Analysis

7.1 Introduction

In this chapter, we consider the situation where there are multiple historical studies for estimating the effect size of the standard therapy as compared to placebo. Many concepts in the analyses of noninferiority (NI) trials are a lot simpler when there is only one historical study than when there are multiple historical studies. For example, what does the constancy assumption mean when there are multiple historical studies? However, multiple historical studies are needed to assess assay sensitivity, which depends on the historical evidence of sensitivity to drug effects (ICH E10 2001).

There are limitations in the meta-analysis to estimate the effect size of the standard therapy as compared to placebo, such as publication bias and heterogeneity. Section 7.2 discusses meta-analysis in general and the associated issues. Section 7.3 discusses the fixed-effect model, and Section 7.4 discusses the use of the random-effects model to deal with heterogeneity. The constancy assumption in the context of meta-analysis will be discussed in Section 7.5. Ng and Valappil (2011) proposed an alternative to deal with heterogeneity by discounting the historical studies individually before pooling. This approach will be discussed in Section 7.6. Finally, Section 7.7 concludes this chapter with a discussion.

7.2 Meta-Analysis and Systematic Review

The definition and objective of a meta-analysis are given by Iyengar and Greenhouse (1988) in the following:

> The application of statistical procedures to collections of results from individual studies for the purpose of integrating, synthesizing, and advancing a research domain is commonly known as meta-analysis. The objective of a meta-analysis is to summarize quantitatively research

literature with respect to a particular question and to examine systematically the manner in which a collection of studies contributes to knowledge about that question.

There are variations in the definition of meta-analysis in the literature. Some examples are given in the following:

- DerSimonian and Laird (1986): Meta-analysis is defined as the statistical analysis of a collection of analytical results for the purpose of integrating the findings.
- Follmann and Proschan (1999): Meta-analysis is an important tool used in medical research to quantitatively summarize multiple related studies.
- Ziegler, Koch, and Victor (2001): Meta-analysis is the systematic synthesis of the results of several studies, especially of clinical trials.
- Schumi and Wittes (2011): Meta-analysis is a set of methods used to combine data from a group of studies to obtain an estimate of a treatment effect.

The studies included in the meta-analysis are typically derived from a systematic review. The Center for Outcomes Research and Education (CORE) contrasts the difference between a systematic review and a meta-analysis in the following (http://researchcore.org/faq/answers.php?recID=5; Accessed: September 7, 2013):

> A **systematic review** is a thorough, comprehensive, and explicit way of interrogating the medical literature. It typically involves several steps, including (1) asking an answerable question (often the most difficult step); (2) identifying one or more databases to search; (3) developing an explicit search strategy; (4) selecting titles, abstracts, and manuscripts based on explicit inclusion and exclusion criteria; and (5) abstracting data in a standardized format.

A "meta-analysis" is a statistical approach to combine the data derived from a systematic review. Therefore, every meta-analysis should be based on an underlying systematic review, but not every systematic review leads to a meta-analysis. Bartolucci and Hillegass (2010, p. 17) elaborate the basic principle of a systematic review contrasting with an informal review as follows:

> The systematic review [follows] an explicit and reproducible protocol to locate and evaluate the available data. The collection, abstraction, and compilation of the data follow a rigorous and prospectively defined objective process. ... Unlike an informal review of the literature, this systematic, disciplined approach is intended to reduce the potential for subjectivity or bias in the subsequent findings.

In addition to published data, O'Gorman et al. (2013) include unpublished data, conference proceedings, and abstracts in a systematic review. Furthermore, the meta-analysis, if performed, is considered part of the systematic review. Khan et al. (2003) explicitly include the meta-analysis as part of the systematic review.

7.3 Fixed-Effect Model

Prior to the introduction of the random-effects model (see Section 7.4) by DerSimonian and Laird in 1986, meta-analysis was mainly based on the fixed-effect model (FEM), assuming there is one true effect size that is shared by all studies. The combined effect size is the estimate of this common effect size (Muthukumarana and Tiwari 2012). For a continuous endpoint under the normality assumption, the common effect size is estimated by the weighted average of the estimates of the individual studies, with weight being the inverse of the estimated variance of individual estimates. See, for example, Rothmann, Wiens, and Chan (2012, p. 75). Without the normality assumption of the underlying distribution, the analysis relies on the asymptotic results on the sample sizes, regardless of the number of studies.

The lower confidence limit for the common effect size should be used to account for the variability in determining the NI margin, with or without discounting for the fixed-margin method discussed in Section 5.3 of Chapter 5. For the synthesis method discussed in Section 5.4 of Chapter 5, the point estimate, with or without discounting, may be incorporated into the test statistic, similar to the situation where there is only one historical study.

7.4 Random-Effects Model

Even when investigating the same disease and the same therapeutic intervention, different studies are almost never identical in design; they might, for example, differ with respect to (1) dosage scheme, (2) duration of follow-up, (3) diagnostic strategies, or (4) the risk profile of the patient population (Ziegler, Koch, and Victor 2001). Such differences could lead to different effect sizes, rendering the assumption for FEM invalid. DerSimonian and Laird (1986) introduced a random-effects model (REM) to take into consideration the heterogeneity between the studies. Under this model, the true effect size could vary across studies, and is assumed to follow a normal distribution. In addition, the between-study and within-study variabilities are assumed to be independent.

As in the FEM, the overall effect size is estimated by the weighted means. See, for example, Rothmann, Wiens, and Chan (2012, 77). The studies in the meta-analysis (see Section 7.2) are assumed to be a random sample of the relevant distribution of effects, and the combined effect estimates the mean effect of this distribution (Muthukumarana and Tiwari 2012). This mean effect is referred to as the global mean by Rothmann, Wiens, and Chan (2012, 61).

Testing under either an FEM or REM typically depends on an asymptotic approximation of a test statistic to a standard normal distribution. For the FEM, the approximation is asymptotic on the total number of subjects, whereas for the REM, it is asymptotic on the number of studies (Follmann and Proschan 1999). Results under the REM based on an asymptotic approximation with a small number of studies are not reliable and are subject to non-negligible bias. DerSimonian and Laird (1986) described the main difficulty in meta-analysis as follows:

> The main difficulty in integrating the results from various studies stems from the sometimes-diverse nature of the studies, both in terms of design and methods employed. Some are carefully controlled randomized experiments while others are less well controlled. Because of differing sample sizes and patient populations, each study has a different level of sampling error as well. Thus, one problem in combining studies for integrative purposes is the assignment of weights that reflect the relative "value" of the information provided in a study. A more difficult issue in combining evidence is that one may be using incommensurable studies to answer the same question.

Rothmann, Wiens, and Chan (2012, 81–82) discussed various concerns with the application of the REM to estimate the effect size in the setting of the NI trial. These are summarized as follows:

- "… the estimation of the mean effect across studies (the global mean) is frequently used to infer the effect of the active control in the NI trial from either the use of a 95% confidence interval or by the use of the point estimate with its standard error (e.g., in a synthesis test statistic)."

- "The lower limit of a two-sided 95% confidence interval for the global mean may increase as the largest estimated effect is reduced," as shown in Example 4.4 of Rothmann, Wiens, and Chan (2012, 82–85), which is counterintuitive. In fact, it is not appropriate to use the global mean, assuming it is known, as the effect size to establish an NI margin, especially since the within-study effects are greatly heterogeneous.

- Including small studies in the systematic review tends to overestimate the effect size in the meta-analysis, as small studies with a smaller observed effect size (and mostly negative results) are less likely to be published. "This bias would be more profound in a random-effects meta-analysis than in a fixed-effect meta-analysis. A random-effects meta-analysis pulls the estimated effect away from that of a fixed-effect meta-analysis toward the observed effects from the smaller studies."

Using an extreme hypothetical example with an infinite number of historical trials, each of infinite size, Brittain, Fay, and Follmann (2012) showed that the global mean (overall mean) should not be used as the effect size in the NI trial. Note that in such an extreme hypothetical example, it effectively assumes that global mean is known. They propose using a prediction interval for the missing standard-versus-placebo effect rather than a confidence interval for the mean. Use of the predictive interval is also discussed by Rothmann, Wiens, and Chan (2012, 108–109).

7.5 Constancy Assumption in the Context of Meta-Analysis

For the FEM, the constancy assumption means that the common effect size is equal to the effect size in the current NI trial, which is similar to the situation where there is only one historical study (see Section 2.5.1 in Chapter 2). Since studies may have different designs, assessing the constancy assumption can be even more challenging than the situation where there is only one historical study.

When the true effect of the active control varies across previous trials (i.e., the active control effect is not constant), such as in the REM, what does the constancy assumption mean? This question was raised by Rothmann, Wiens, and Chan (2012, 61). They gave two interpretations: (1) the active control effect in the NI trial equals the global mean active-control effect across studies (i.e., the location parameter in the REM); and (2) the true active-control effect in the NI trial has the same distribution as the true effects in the previous trials.

Whether to use the confidence interval for the global mean or the predictive interval for estimating the effect size in the current NI trial depends on the interpretation of the constancy assumption. With the first interpretation, we should use the confidence interval for the global mean, and discounting may be used if the constancy assumption does not hold, as in Section 2.5.2 of Chapter 2.

However, Brittain, Fay, and Follmann (2012) are against using the global mean even if it is known, as discussed in Section 7.4. With the second interpretation, we may consider the studies in the meta-analysis as a random sample of the relevant distribution of effects (Muthukumarana and Tiwari 2012) (see Section 7.4), with the effect size of the current NI trial being missing; therefore, we should use the predictive interval rather than the confidence interval. In this case, it is not clear what is meant by violation of the constancy assumption because there is no explicit formula to equate the effect size in the current NI trial and the "effect size" of the historical studies. Perhaps, violation of the constancy assumption means that the assumption of the REM is not valid. In any case, the estimated effect size using the predictive interval can always be discounted. Therefore, the effect size in the current NI trial is obtained by first pooling the data from the historical studies using the predictive interval and then applying the discounting, if needed. Such an approach will be referred to as the Pooling-and-then-Discounting (PatD) approach.

It is not clear how realistic the assumption for the second interpretation is. Even if this assumption holds, the number of studies is typically limited.

For example, DerSimonian and Laird (1986) described seven medical meta-analyses, six of which had fewer than 10 studies. With a limited number of studies, Ng and Valappil (2011) proposed discounting each individual study and then pooling the results. This approach will be discussed in Section 7.6.

When we use the confidence interval for the global mean or the predictive interval for estimating the effect size in the current NI trial, we implicitly use the fixed-margin method discussed in Section 5.3 of Chapter 5. Using the synthesis method with the first interpretation, we incorporate the point estimate for the global mean in the test statistic, as shown in Section 5.4 of Chapter 5. Using the synthesis method with the second interpretation, we incorporate the point estimate for a given percentile (e.g., 10th percentile) of the distribution of the truth means in the REM, rather than the global mean in the test statistic, as shown in Section 5.4 of Chapter 5. In either case, the variability of the test statistic needs to be accounted for appropriately.

7.6 Discounting-and-then-Pooling Approach

In both FEM and REM analyses, a discounting of the estimated effect size in the meta-analysis may be considered. Since the discounting is done after the data is pooled, it is referred to as the Pooling-and-then-Discounting (PatD) approach (see Section 7.5). Ng and Valappil (2011) proposed the Discounting-and-then-Pooling (DatP) approach when the number of historical studies is small (e.g., three). A simple hypothetical example will be discussed in Section 7.6.1 to illustrate the DatP approach. In Section 7.6.2, an example based on published literature in the anti-infective therapeutic area is used to contrast the two approaches.

The focus is on estimating the effect size in the current NI trial using data from the historical studies, that is, $(S - P)$, the mean difference for the continuous outcomes or the difference in proportions for the binary outcomes. Typically, the lower limit of the associated 95% confidence interval is used for such an estimate. This estimate may be subjected to discounting, depending upon the validity of the constancy assumption (see Section 2.5 of Chapter 2) and is used to determine the NI margin for the fixed-margin method (see Section 5.3 of Chapter 5). Note that the notation "d" is used in this section instead of "γ" to avoid confusing the three components (i.e., γ_1, γ_2, and γ_3) discussed in Section 2.5.2 of Chapter 2.

7.6.1 A Simple Hypothetical Example

Suppose there are three previous trials. Let

$$(S - P)_c = d_1 (S - P)_{h_1}$$
$$(S - P)_c = d_2 (S - P)_{h_2}$$
$$(S - P)_c = d_3 (S - P)_{h_3}$$

TABLE 7.1

How to Choose d_1, d_2 and d_3

Study	Time the Study Was Conducted	Discounting	d_i
1	5 years ago	10%	0.9
2	10 years ago	20%	0.8
3	15 years ago	30%	0.7

where $0 \le d_i \le 1$, for $i = 1, 2, 3$ and d_i's are the rates of discounting. Accordingly, these studies are discounted by $(1 - d_1)100\%$, $(1 - d_2)100\%$, and $(1 - d_3)100\%$, respectively. After discounting, we then pool the three studies by taking the average, as given in the following:

$$(S - P)_c = [d_1(S - P)_{h_1} + d_2(S - P)_{h_2} + d_3(S - P)_{h_3}] / 3$$

So, the effect size is the average of the effects of the individual studies after discounting. The effect size may be estimated by the weighted average (see Section 7.3) of the individual estimates with discounting.

How do we choose d_1, d_2, and d_3? We consider a simplistic situation for illustration purposes. Let us assume that the three studies were identical in every aspect, except for the time when the studies were conducted. Suppose that studies 1, 2, and 3 were conducted 5, 10, and 15 years ago (see Table 7.1). If we decide to discount study 1 by 10% (so, $d_1 = 0.9$), then we would proportionally discount studies 2 and 3 by 20% and 30%, respectively. Note that determination of discounting for study 1 is subjective. However, once that is decided, a logical and simple way is to discount the other two studies proportionally.

When two or more factors might affect the constancy assumption, each factor can be assessed individually, and the composite discounting may then be calculated. This will be shown in the example in the next sub-section.

7.6.2 Anti-Infective Example

This example contrasts two approaches that deal with discounting to determine the effect size of an NI trial for *Clostridium difficile* infection that induces severe diarrhea in patients compromised by antibiotic usage and other underlying disease conditions. Vancomycin is chosen as the active control in the current NI trial example. The study objective is to show the efficacy of an experimental treatment as compared to putative placebo (i.e., $\varepsilon = 1$; see Section 2.4 in Chapter 2); therefore, the NI margin is determined using the estimated effect size in the NI trial with discounting.

There is limited information on placebo-controlled trials using vancomycin in the literature. Therefore, in this example, two large, phase 3, randomized, multicenter, double-blind, controlled studies—referred to as studies 301 and 302—comparing vancomycin to tolevamer were used to estimate the effect of vancomycin treatment over tolevamer (Weiss 2009). These studies were completed during March 2005 through August 2007 and were originally designed to demonstrate that tolevamer is noninferior to vancomycin using an NI margin of 15%, but tolevamer was found to be inferior to vancomycin, and further development of tolevamer was stopped (Weiss 2009). In this example, tolevamer was assumed to be no worse than placebo, and was considered as a placebo in determining the effect size for designing future NI trials.

These two studies (301 and 302) utilized vancomycin 125 mg four times a day (q.i.d.) for 10 days and tolevamer 3 g dosing three times a day (t.i.d.) for 14 days. An initial 9 g loading dose was utilized among the tolevamer treated patients. The results of these studies have already been published or presented at conferences (e.g., Bouza et al. 2008; Louie et al. 2007; U.S. FDA 2011; Optimer 2011). These studies were originally designed as phase 3, multicenter, randomized, double-blind, parallel studies with patients enrolled from the United States, Canada, Europe, or Australia. Patients were randomized (2:1:1) to tolevamer (3 g t.i.d., 14 days), vancomycin (125 mg q.i.d., 10 days), or metronidazole (375 mg q.i.d., 10 days). Note that the metronidazole arm is not used in this example.

Clinical success was used as the outcome in those studies and was defined as resolution of diarrhea and absence of severe abdominal discomfort due to *Clostridium difficile*—associated diarrhea (CDAD) on day 10. This example considers two different metrics to evaluate the treatment effect—namely, risk difference and odds ratio. The estimated individual treatment effects, as well as the pooled treatment effect, with the associated 95% confidence intervals are summarized in Table 7.2.

Comparing studies 301 and 302 to the recent NI trials, it appears that the strains of C. *difficile* and susceptible populations are likely to be similar. However, there can potentially be few differences, compared to recent trials in CDAD, with respect to: (1) entry criteria, (2) definitions of clinical success, (3) emerging resistance on vancomycin, and (4) other factors.

The proportions of patients with severe symptoms varied and were different between historical and current NI trials. Patients enrolled in Study 301 had symptoms that were more severe than those enrolled in Study 302, in terms of percent of patients with severe disease at baseline (32% versus 25%).

Study 301 reported 37% dropout rate where the dropouts included non-response, death, lost to follow-up, voluntary withdrawal, etc. (Louie et al. 2007; U.S. FDA 2011; Optimer 2011). The dropout rate in Study 302 was not available. It is assumed to be 44% in this example to illustrate the method. Recent NI trials have enrolled patients with baseline disease severity in the range of 35%–39% with dropout rates less than 10% (see Table 7.3). For illustration of the DatP approach, disease severity (in terms of the proportion of patients with severe symptoms) and dropout rate are used in the determination of discounting.

For the PatD approach with a 20% discounting, the effect size in the current NI trial is estimated at 0.8 × (lower limit of the 95% confidence interval), which is equal to (1) $\delta = 0.8 \times 30.5\% = 24.4\%$, for the difference metric (i.e.,

TABLE 7.2

Clinical Success Rates: Intent-to-Treat Analysis

Study	Clinical Success Rate		Difference (S – P) (95% CI)	Odds Ratio [S/(1 – S)]/ [P/(1 –P)] (95% CI)
	Tolevamer (P)[1] x/n (%)	Vancomycin (S) x/n (%)		
Study 301 (Louie et al. 2007)	124/266 (46.4)	109/134 (80.7)	34.7% (25.8%, 43.6%)	4.99 (3.04, 8.21)
Study 302 (Bouza et al. 2008)	112/268 (41.6)	101/125 (80.2)	39.0% (29.9%, 48.1%)	5.86 (3.53, 9.73)
Weighted Meta-Analysis[2]			36.8% (30.5%, 43.2%)	5.40 (3.79, 7.70)

[1] Tolevamer is used as placebo.
[2] The meta-analysis used a random-effects model based on the DerSimonian-Laird approach implemented by Comprehensive Meta-Analysis version 2.2.064.

TABLE 7.3

Patient Severity and Dropout Rates: Intent-to-Treat Analysis

Study	Disease severity (%)[1]	Dropout rate (%)
Study 301 (Louie et al. 2007)	32%	37%[1]
Study 302 (Bouza et al. 2008)	25%	44%[2]
Recent NI trials (U.S. FDA 2011; Optimer 2011)	35%–39%	<10%

[1] Combined estimate for tolevamer and vancomycin.
[2] The dropout rate is assumed to be 44% for illustration purposes.

S – P; see Section 4.2 in Chapter 4); and (2) $r = 3.79^{0.8} = 2.90$, for the odds ratio metric (see Section 4.3 in Chapter 4).

An alternative approach is to discount each individual study first and then pool (DatP). To illustrate this approach, each individual study was separately discounted based on disease severity and dropout rate, and the overall discounting was then calculated, as shown in Table 7.4. For example, assuming disease severity of about 39% (see Table 7.3) for the current NI trials and using a 5% discounting for study 301, study 302 should then be discounted by 10% (= 5% × 14%/7%). Similarly, assuming a dropout rate of 10% (see Table 7.3) for the current NI trials and using a 6% discounting for study 301, study 302 should then be discounted by 7.6% (= 6% × 34%/27%). The overall discounting for studies 301 and 302 is 10.7% and 16.84%, respectively, as shown in Table 7.4. Using the lower limit of the 95% confidence interval for the absolute difference in each study in Table 7.2, the effect size in the current NI trial is estimated at (1) $\delta = 0.5[0.8930(25.8\%) + 0.8316(29.9\%)] = 24.0\%$, for the difference metric (i.e., S – P; see Section 4.2 in Chapter 4); and (2) $r = (3.04^{0.8930} \times 3.53^{0.8316})^{0.5} = 2.76$, for the odds ratio metric (see Section 4.3 in Chapter 4).

TABLE 7.4

Discounting Using DatP Approach

Study	Disease Severity (%)	Discounting (%)	Dropout Rate (%)	Discounting (%)	Overall Discounting
Study 301 (Louie et al. 2007)	32%	5%	37%	6%	$1 - (0.95 \times 0.94)$ $= 10.70\%$
Study 302 (Bouza et al. 2008)	25%	10%	44%	7.6%	$1 - (0.90 \times 0.924)$ $= 16.84\%$

Calculations of the NI margins δ and r, with the objective to show efficacy (i.e., ε = 1; see Section 2.4 in Chapter 2) using the PatD and DatP approaches for discounting are summarized in Table 7.5. It should be noted that showing efficacy of an anti-infective agent is, in general, not sufficient for approval by the FDA. The study should be designed to show that certain percent of the control effect be preserved based on a clinical judgment. In this example, if the study objective is to show greater than 50% preservation using the difference as the metric, then the NI margins are 12.2% and 12% using the PatD and DatP methods, respectively. However, the clinically acceptable NI margin for this type of bacterial infection trials should be set to 10%. In other words, the NI margin should not be larger than 10 percentage points.

TABLE 7.5

PatD and DatP Approaches for Discounting

Study	Discounting: (1 – d)	Difference: S – P		Odds Ratio: S/(1 – S)]/[P/(1 – P)	
		Lower Limit of the 95% CI[1]	NI Margin[2]: δ	Lower Limit of the 95% CI[1]	NI Margin[3]: r
Pooling-and-then-Discounting approach (PatD)					
Weighted Meta-Analysis[2]	20%	30.5%	24.4%	3.79	2.90
Discounting-and-then-Pooling approach (DatP)					
Study 301 (Louie et al. 2007)	10.70%	25.8%	23.04%	3.04	2.79
Study 302 (Bouza et al. 2008)	16.84%	29.9%	24.86%	3.53	2.85
Pooling[4]	—	—	24.0%	—	2.76

[1] From Table 7.2
[2] $\delta = d \times$ (lower limit of the 95% confidence interval)
[3] $r =$ (lower limit of the 95% confidence interval)$^{1-d}$
[4] Arithmetic mean (i.e., average) for δ and geometric mean for r

7.7 Discussion

Meta-analysis has its limitations. In addition to the concerns raised by Rothmann, Wiens, and Chan (2012, 81–82) (see Section 7.4), two well-known major problems with the meta-analysis are given by Hung, Wang, and O'Neill (2009): (1) publication bias due to the fact that negative studies are rarely published in the literature, and (2) how to weigh each study in the meta-analysis. Exclusion of negative studies may overestimate the treatment effect. Using invalid FEM (See Section 7.3) in the meta-analysis can seriously overestimate the variance of the treatment effect estimate. On the other hand, the REM (See Section 7.4) may give similar weights to all studies regardless of sample size, which is undesirable when the sample sizes varies greatly. (Hung, Wang, and O'Neill 2009).

The DatP approach proposed in Section 7.6 provides an alternative, especially when the number of historical studies is small. Although the discussion of the DatP approach in Section 7.6.2 focuses on the fixed-margin method, the synthesis method (see Section 5.4 in Chapter 5) may be used, but is omitted here.

References

Bartolucci AA and Hillegass WB (2010). Overview, Strengths, and Limitations of Systematic Reviews and Meta-Analyses Overview, in *Evidence-Based Practice: Toward Optimizing Clinical Outcomes*, eds. Chiappelli F, Caldeira Brant XM, Neagos N, Oluwadara OO, and Ramchandani, MH. Berlin Heidelberg: Springer.

Bouza E, Dryden M, Mohammed R, Peppe J, Chasan-Taber S, Donovan J, Davidson D, and Short G (2008). Results of a Phase III Trial Comparing Tolevamer, Vancomycin and Metronidazole in Patients with *Clostridium Difficile*-Associated Diarrhea. Poster Abstract Number: O464. 18th European Congress of Clinical Microbiology and Infectious Diseases Barcelona, Spain, April 19–22, 2008.

Brittain EH, Fay MP, and Follmann DA (2012). A Valid Formulation of the Analysis of Noninferiority Trials Under Random-Effects Meta-analysis. *Biostatistics*, **13**(4):637–649.

DerSimonian R and Laird N (1986). Meta-analysis in Clinical Trials. *Controlled Clinical Trials*, **7**:177–188.

Follmann DA and Proschan MA (1999). Valid Inference in Random-Effects Meta-Analysis. *Biometrics*, **55**:732–737.

Hung HMJ, Wang S-J, and O'Neill R (2009). Challenges and Regulatory Experiences with Non-Inferiority Trial Design Without Placebo Arm. *Biometrical Journal*, **51**:324–334.

International Conference on Harmonization (ICH) E10 Guideline (2001). *Choice of Control Groups in Clinical Trials*. http://www.fda.gov/downloads/Drugs/GuidanceComplianceRegulatoryInformation/Guidances/UCM073139.pdf (Accessed: September 27, 2012).

Iyengar S and Greenhouse JB (1988). Selection Models and the File Drawer Problem. *Statistical Science*, 3:109–135.

Khan KS, Kunz R, Kleijnen J, and Antes G (2003). Five Steps to Conducting a Systematic Review. *Journal of the Royal Society of Medicine*, **96**:118–121.

Louie TJ, Gerson M, Grimard D, Johnson S, Poirier A, Weiss K et al. (2007). Results of a Phase III Study Comparing Tolevamer, Vancomycin and Metronidazole in *Clostridium Difficile*–Associated Diarrhea (CDAD), in Program and Abstracts of the 47th Interscience Conference on Antimicrobial Agents and Chemotherapy (ICAAC); September 17–20, 2007; Chicago, IL. Washington DC: ASM Press; Abstract K-4259.

Muthukumarana S and Tiwari RC (2012). Meta-analysis Using Dirichlet Process. *Statistical Methods in Medical Research*. http://smm.sagepub.com/content/early /2012/07/16/ (Accessed: July 24, 2012).

Ng T-H and Valappil T (2011). Discounting and Pooling of Historical Data in Noninferiority Clinical Trials. Unpublished manuscript.

O'Gorman CS, Macken AP, Cullen W, Saunders J, Dunne C, and Higgins MF (2013). What Are the Differences between a Literature Search, a Literature Review, a Systematic Review, and a Meta-analysis? And Why Is a Systematic Review Considered to Be So Good? *Irish Medical Journal*, **106**(2):8–10. http://ulir.ul.ie /handle/10344/3011 (Accessed: September 7, 2013).

Optimer Pharmaceuticals, Inc. (2011). Dificid™ (Fidaxomicin Tablets) for the Treatment of *Clostridium Difficile* Infection (CDI), Also Known as *Clostridium Difficile*–Associated Diarrhea (CDAD), and for Reducing the Risk of Recurrence when Used for Treatment of Initial CDI, NDA 201699: Anti-Infective Drugs Advisory Committee Meeting Briefing Document, April 5, 2011. http://www .fda.gov/downloads/AdvisoryCommittees/CommitteesMeetingMaterials /Drugs/Anti-InfectiveDrugsAdvisoryCommittee/UCM249354.pdf (Accessed: August 16, 2013).

Rothmann MD, Wiens BL, and Chan ISF (2012). *Design and Analysis of Non-Inferiority Trials*. Boca Raton, FL: Chapman & Hall/CRC.

Schumi J and Wittes JT (2011). Through the Looking Glass: Understanding Non-inferiority. *Trials*, **12**:106. http://www.trialsjournal.com/content/12/1/106 (Accessed: August 25, 2013).

U.S. Food and Drug Administration (2011). Fidaxomicin for the Treatment of *Clostridium Difficile*-Associated Diarrhea (CDAD): FDA Briefing Document for Anti-Infective Drugs Advisory Committee Meeting, April 5, 2011. http://www .fda.gov/downloads/AdvisoryCommittees/CommitteesMeetingMaterials /Drugs/Anti-InfectiveDrugsAdvisoryCommittee/UCM249353.pdf (Accessed: August 16, 2013).

Weiss K (2009). Toxin-Binding Treatment for *Clostridium difficile*: A Review Including Reports of Studies with Tolevamer. *International Journal of Antimicrobial Agents*, **33**:4–7.

Ziegler S, Koch A, and Victor N (2001). Deficits and Remedy of the Standard Random-Effects Methods in Meta-analysis. *Methods of Information in Medicine*, **40**:148–155.

8

Three Treatment Groups

8.1 Introduction

So far, we have been dealing with two-arm trials comparing a test treatment with either an active control or the standard therapy. In this chapter, we consider multiple-arm clinical trials with three treatment groups with continuous endpoints. We assume that the underlying distribution is normal and use the same notations where applicable (e.g., T, S, and P), but with modifications where needed (e.g., T_1, T_2, and T_3 denote test treatments 1, 2, and 3, respectively). We assume that a larger value corresponds to a better outcome.

A three-arm trial could be one of the following: (1) comparing a test treatment (T) with an active control (or standard therapy; S) and a placebo (P), (2) comparing two test groups (T_1 and T_2) with an active control (S), and (3) testing equivalence of three treatment groups (T_1, T_2, and T_3) without a control, such as lot consistency or lot release studies. These will be discussed in Sections 8.2, 8.3, and 8.4, respectively.

8.2 Gold-Standard Design

8.2.1 Reasons for the Gold-Standard Design

Although a randomized, double-blind, placebo-controlled trial is the gold standard in assessing the efficacy of the test treatment (see Section 1.5.1 of Chapter 1), when an effective treatment exists, there is a consensus that a three-arm trial, including a test treatment (T), an active control (or standard therapy; S), and a placebo (P), is the gold standard (referred to as STP) if placebo use is ethical (e.g., Koch and Rohmel 2004; Hauschke and Pigeot 2005). The STP design may evolve from either a NI trial by adding a placebo arm or a placebo-control trial by adding an active control arm, assuming placebo use is ethical.

Adding a placebo arm to a noninferiority (NI) trial is often recommended in situations where assay sensitivity cannot be established, such as in studies of antidepressant drugs. Such a recommendation is warranted only if the study objective of the NI trial is to show comparative effectiveness. If the study objective of the NI trial is simply showing the efficacy of the test treatment as compared to placebo, a placebo-controlled trial rather than the NI trial should be conducted.

The U.S. Food and Drug Administration (FDA) (2010) draft guidance on NI trials states that "where comparative effectiveness is the principal interest, it is usually important—where it is ethical, as would be the case in most symptomatic conditions— to include a placebo control as well as the active control." The European Medicines Agency (EMA) (2005) guideline on the choice of the NI margin recommends such a design when it states that "a three-armed trial with test, reference, and placebo allows some within-trial validation of the choice of [NI] margin and is therefore the recommended design; it should be used wherever possible."

Koch and Rohmel (2004) presented situations where it is wise to include an additional placebo group instead of just performing a two-arm NI clinical trial:

- Reference is a "traditional" standard (i.e., an established treatment for which principal proof of efficacy is lacking and doubts in this efficacy exist, or an established treatment [was] tested long ago [and] the relevance of the historical finding in the present medical setting is unclear).

- Reference is a "weak" standard (i.e., the difference between reference and placebo is small and it might be difficult to justify a negligible loss of efficacy δ).

- Reference is a "volatile" standard (i.e., in different trials, different estimates for the treatment effect as compared to placebo have been observed, and no accepted explanation for these differences is available).

- The disease under investigation is not fully understood (i.e., not only response to reference, but also response to placebo, varies without constancy in the treatment effect).

The authors also presented advantages, especially in the regulatory setting, of including the reference treatment group in a two-arm, placebo-controlled clinical trial:

- Placebo comparisons may be meaningless where a well-established reference that might seem to outperform the experimental treatment exists.

- [If] the experimental treatment does not work in a certain clinical trial as compared to placebo, this failed study may cause problems with the application for the drug license. For example, recent guidance on meta-analysis (CPMP 2000) explicitly requests that at least positive trends be observed in all studies to be combined. A negative study always weakens the evidence for the efficacy of the experimental treatment. Here, it would be helpful to balance the fact that the experimental treatment failed with the knowledge that the established standard also did not work.

- Rare situations may exist in which superiority of the experimental treatment over placebo alone is not convincing without the additional evidence that the reference was also superior to placebo in the same trial. This, for example, is one interpretation of the current Committee for Proprietary Medicinal Products (CPMP) guidance regarding approval of new antidepressants (CPMP 2002a).

The International Conference on Harmonization (ICH) E10 (2001, pp. 13–14) elaborates on the purpose of (1) adding an active-control arm to a placebo-controlled trial or (2) adding an active-control arm to a placebo-controlled trial to assess assay sensitivity:

> The question of assay sensitivity, although particularly critical in [NI] trials, actually arises in any trial that fails to detect a difference between treatments, including a placebo-controlled trial and a dose-response trial. If a treatment fails to show superiority to placebo, for example, it means either that the treatment was ineffective or that the study as designed and conducted was not capable of distinguishing an effective treatment from placebo.
>
> A useful approach to the assessment of assay sensitivity in active-control trials and in placebo-controlled trials is the *three-arm trial*, including both placebo and a known active treatment, a trial design with several advantages. Such a trial measures effect size (test drug versus placebo) and allows comparison of test drug and active control in a setting where assay sensitivity is established by the active control versus placebo comparison (see Section 2.1.5.1.1).

The choice of study design, such as two-arm, placebo-controlled (TP); two-arm, active-control (ST); or three-arm with active and placebo controls (STP), in the evaluation of a test treatment depends on (1) the existence of an effective treatment, (2) the study objective, (3) ethical use of placebo, and (4) assay sensitivity (see Section 2.6 in Chapter 2 for the definition of assay sensitivity). If no effective treatment for a given disease exists, the two-arm, placebo-controlled trial is the only choice. Assuming an effective treatment exists, there are two possible objectives in assessing a test treatment: Objective 1 (O1) assesses the efficacy of the test treatment as compared to placebo, and Objective 2 (O2) assesses the efficacy of the test treatment relative to the active control (or standard therapy).

TABLE 8.1

Choice of Study Design Assuming an Effective Treatment Exists

Study objective:	Placebo use ethical	Assay sensitivity	Study design*
O1[1] Efficacy	Yes	—	TP
	No	Yes	ST
		No	Other (e.g., add-on)
O2[2] Comparative efficacy	Yes	—	STP
	No	Yes	ST
		No	None

* TP: two-arm, placebo-controlled; ST: two-arm, active-control; STP: three-arm trial.
[1] O1: To assess the efficacy of the test treatment as compared to placebo.
[2] O2: To assess the efficacy of the test treatment relative to the active control (or standard therapy).

For O1, if placebo use is ethical, then the two-arm, placebo-controlled trial (TP) should be used, regardless of assay sensitivity; otherwise, the two-arm, active-control trial (ST) may be used, provided assay sensitivity can be established. For O2, if placebo use is ethical, then the three-arm trial (STP) should be used, regardless of assay sensitivity; otherwise, the two-arm, active-control trial (ST) may be used, provided assay sensitivity can be established. If placebo use is unethical and assay sensitivity cannot be established, then other designs (e.g., add-on design) should be considered for O1, and no design may be used for O2 (see Table 8.1).

8.2.2 Controversial Issues

The STP design is more complex than the two-arm trial, since the three pairwise comparisons (T versus P, T versus S, and S versus P) may be of interest. Whether or not the superiority of S to P would be a mandatory prerequisite for interpretation of the trial is controversial. The statement regarding establishment of assay sensitivity in ICH E10 (see Section 8.2.1) appears to say that the superiority of S to P would be a mandatory prerequisite. The traditional strategy is to simultaneously test the following three null hypotheses at the prespecified significance level α, which is typically set to 0.025 (Koch and Rohmel 2004):

- H_{01}: $T \leq P$ against H_{a1}: $T > P$, for showing efficacy of T as compared to P
- H_{02}: $T \leq S - \delta$ against H_{a2}: $T \leq S - \delta$, for showing NI of T relative to S
- H_{03}: $S \leq P$ against H_{a3}: $S \leq P$, for showing efficacy of S as compared to P

The trial would be considered successful only if all three null hypotheses are rejected. Pigeot et al. (2003) suggested testing the NI hypothesis (i.e., H_{02})

at the significance level α only if superiority of S over P has been shown by rejecting the null hypothesis that $S \le P$ (i.e., H_{03}) at the same α level. There is no Type I error adjustment, as the hypotheses are tested in a hierarchical order. However, Koch and Rohmel (2004) objected to such hierarchical order and argued that such a mandatory requirement of showing $S > P$ is ill founded for the following reasons:

- Given that the experimental is noninferior to the reference and the reference is superior to placebo, the trial would not be accepted as proof of efficacy if, in this situation, superiority of the experimental treatment over placebo was not established. Consequently, superiority of the experimental treatment over placebo is a mandatory prerequisite in this setting. Under this condition, however, assay sensitivity is proven, and the superiority of the reference over placebo is not needed to demonstrate assay sensitivity.

- Doubt in the reference treatment's ability to discern from placebo (whether because of limited knowledge, small distance with respect to response rates, or difficulties in providing a credible estimate of the reference response) has been a major reason for including a placebo group in an active-control trial. The reference treatment that fails to demonstrate superiority over placebo under these prerequisites, and at the same time an experimental treatment that successfully discerns from placebo, should be seen as an additional strength of the experimental treatment.

- It might be deemed necessary to also establish the superiority of the reference treatment over placebo to prove that "the correct patient population" has been identified in a certain trial. The rationale for this argument is again unclear because in later clinical practice, no pretesting is done in order to identify those patients who will benefit from the experimental treatment. Trials that fail because of incorrect estimation of the placebo response (as in depression) might be the result of the inability to successfully describe a homogeneous patient population for the trials. There is no good reason to doubt the efficacy of the experimental treatment only because of the reference treatment's inability to discern from placebo.

The authors proposed an assessment of efficacy in a step-down procedure to control the family-wise error rate, which is outlined as follows:

Step 1: Test H_{01}; if it can be rejected at a prespecified level α, then go to step 2; otherwise, stop.

Step 2: Test H_{02}; if it can be rejected at the same level α, then go to step 3; otherwise, stop.

Step 3: Test H_{03} and H_{04}, simultaneously at the same level α, where:

- H_{04}: $T \leq S$ against H_{a4}: $T > S$, to show superiority of T as compared to S.

If both H_{01} and H_{02} are false, both H_{03} and H_{04} cannot be true simultaneously because $T \leq S$ and $S \leq P$ contradict $T > P$. Therefore, the Type I error is controlled (Shaffer 1986; Hommel 1988). Koch and Rohmel (2004) concluded that "unless further arguments can be provided, regulatory requirements for the assessment of gold-standard [NI] trials should be limited to the demonstration of superiority of the experimental over placebo and the demand that [NI] of the experimental as compared to the reference is demonstrated." However, Hauschke and Pigeot (2005, p. 784) argued that these conditions for a successful trial might not be sufficient from a regulatory point of view. Their argument is as follows with notations used in this book:

> To illustrate our concerns and our argumentation in favor of a gold-standard design, we first assume a medical indication where the reference represents a traditional standard with doubts in efficacy; that is $S \leq P$. This issue is well recognized, for example, in studies of antidepressant drugs, where it might be difficult to distinguish between placebo and the reference (Temple and Ellenberg 2000). Furthermore, let the experimental treatment be superior to placebo; that is $T > P$. [NI] of the new treatment relative to reference $T > S - \delta$ can be concluded for any margin $\delta > 0$. Under these circumstances, we agree that efficacy of the experimental treatment over placebo can be claimed. However, let us now consider the clinical investigation in patients with mild persistent asthma, where a three-arm study, including placebo and a corticosteroid as an active comparator, is strongly recommended by the Note for Guidance on the clinical investigation of medicinal products in the treatment of asthma (CPMP 2002b). Failure to show superiority of the corticosteroid over placebo will challenge the quality of the whole study, with the consequence that even if superiority of the new experimental treatment over placebo can be shown, this might not be accepted by regulatory authorities for a claim of efficacy. Hence, we conclude that assay sensitivity is a mandatory condition whenever a well-established comparator is included in the gold-standard design.

Koch (2005, p. 792) made a strong argument in the following against the requirement of showing superiority of the reference over placebo where (1) the experimental treatment can be shown to be superior to placebo and (2) the experimental treatment can be shown to be noninferior to the reference treatment:

> In support of our argument that, at a minimum, the [previously] mentioned claims (1) and (2) have to be substantiated, we have identified situations, where, although a reference exists, placebo may be needed in addition. [The] main reasons were that reference is a traditional standard with not much or outdated scientific support (e.g., because the co-medication has changed

completely for the disease under investigation), or a weak standard (in that it would be difficult to justify [an NI] margin in an active controlled, two-arm trial). Should, in such a situation, a new experimental treatment that has shown to be superior to placebo and noninferior (or even better than reference) be blamed for the fact that reference could not beat placebo?

Both arguments against and for the requirement to show superiority of the reference over placebo are indisputable. The apparent disagreement is caused by using different reference/disease area in their arguments. For example, in the studies of antidepressant drugs, it might be difficult to distinguish between placebo and the reference; therefore, the reference should not be expected to show superiority over placebo. On the other hand, failure to show superiority of a corticosteroid as an active comparator over placebo in a clinical investigation in patients with mild, persistent asthma might not be accepted by regulatory authorities for a claim of efficacy, even if superiority of the new experimental treatment over placebo can be shown.

In the Koch-Rohmel procedure (Koch and Rohmel 2004), if NI of the experimental treatment cannot be shown, the superiority of the reference over placebo can no longer be assessed within this confirmatory strategy, because confirmatory testing has to stop after nonrejection of H_{02} to control the family-wise error rate in the strong sense. To overcome this disadvantage, Röhmel and Pigeot (2010) modified the Koch-Rohmel procedure by testing H_{02} and H_{03} simultaneously in step 2 rather than testing only H_{02}. Since H_{02} and H_{03} cannot both be true if H_{01} is false, no α-adjustment is needed.

8.2.3 Testing for Noninferiority in the STP Design

In the ST design of an NI trial (see Section 8.2.1), the margin is commonly set as a fraction of the effect size, as proposed in Section 2.2 of Chapter 2 given by Equation 2.1; that is $\delta = \varepsilon(S - P)$, where $(S - P)$ is estimated from the historical data through meta-analysis (see Chapter 7 and Pigeot et al. 2003). Chapter 5 discussed two approaches for testing the NI hypothesis, namely, fixed-margin and synthesis methods. Unlike the fixed-margin method, no margin is actually determined using the synthesis method (see Section 5.4 of Chapter 5). There are two versions of the synthesis approach. The version discussed in Section 5.4 of Chapter 5 is the linearized version. The other version is based on the fraction of effect preserved, defined as the ratio of (1) the effect of the experimental treatment versus placebo in the current NI trial over (2) the effect of standard treatment (active control) versus placebo in the historical data (Hassalblad and Kong 2001). This approach is known as the delta method (e.g., Rothmann and Tsou 2003) and will be referred to as the ratio-based method.

As discussed in Section 5.3, using the fixed margin method in the ST design NI trial, $(S - P)$ is estimated from the historical data in the determination of the NI margin. In the STP design, no such estimate is needed to determine the NI margin, except in the planning stage, where the margin is used in the sample size calculation (Röhmel and Pigeot 2010). Therefore, the fixed-margin approach is no longer applicable in the STP design. Similar to the synthesis approach, by incorporating the NI margin given by Equation 2.1 (assuming $S > P$) into Equations 1.3a and 1.3b, we can test

$$H_0 : T - S \le -\varepsilon(S - P) \tag{8.1a}$$

versus

$$H_1 : T - S > -\varepsilon(S - P) \tag{8.1b}$$

or equivalently,

$$H_0 : T - (1 - \varepsilon)S - \varepsilon P \le 0 \tag{8.2a}$$

versus

$$H_1 : T - (1 - \varepsilon)S - \varepsilon P > 0 \tag{8.2b}$$

These hypotheses were formulated by Ng (1993, 2001) and Pigeot et al. (2003). The null hypothesis may be tested based on the standard t-statistic as given by Pigeot et al. (2003). This is similar to the linearized synthesis method with the ST design (see Section 5.4 in Chapter 5). Alternatively, the hypotheses in Equation 8.1 can be reformulated as the ratio of $(T - P)$ over $(S - P)$, as follows (Pigeot et al. 2003; Koch and Rohmel 2004; Hauschke and Pigeot 2005):

$$H_0 : (T - P)/(S - P) \le 1 - \varepsilon \tag{8.3a}$$

versus

$$H_1 : (T - P)/(S - P) > 1 - \varepsilon \tag{8.3b}$$

The null hypothesis can be tested using the general form of Fieller's theorem, which takes the stochastic dependence of the numerator and denominator into account in constructing the confidence interval for this ratio (Pigeot et al. 2003). This is similar to the ratio-based synthesis method in the ST design, except that the numerator and denominator are independent, as they are estimated from "two" independent studies: one from the current NI trial and the other one from the meta-analysis of the historical studies. Since we are dealing with only the current STP design trial, there is no "synthesis" in the test statistic. We will refer to testing procedure based on Equations 8.2 and 8.3—the linearized approach and the ratio-based approach, respectively—dropping the "synthesis."

Hida and Tango (2011a) considered an STP design trial with a prespecified margin. They argued that it is not appropriate to use $-\varepsilon(S - P)$, as the NI margin must be a prespecified difference in means in the protocol. To test the

NI of the experimental treatment to the reference with the assay sensitivity in this three-arm trial, they proposed that we have to show simultaneously (1) the NI of the experimental treatment to the reference treatment with an NI margin δ (i.e., $T > S - \delta$) and (2) the superiority of the reference treatment to the placebo by more than δ (i.e., $S > P + \delta$). Rohmel and Pigeot (2011, pp. 3162–3163) questioned the requirement to show $S > P + \delta$ in the following:

> As far as we have understood the statement of the authors correctly, "assay sensitivity" can only be stated when the reference shows a "substantial" superiority over placebo ($S - \delta$ superior to P), in contrast to our view that "simple" superiority (S superior to P) is sufficient to demonstrate "assay sensitivity." We are not aware that such a requirement of substantial superiority is widely in use, but we are not familiar with the regulatory demands in Japan. Moreover, we do not see a rationale behind the asymmetry in requirements for substantial superiority for the reference ($S - \delta$ superior to P) and only simple superiority for the experimental therapy (T superior to P).

Hida and Tango (2011b) argued that the requirements of showing "$T > P$," "$S > P$," and "$T > S - \delta$" are not sufficient, because showing NI of T over S with an NI margin of δ (i.e., $T > S - \delta$) is meaningless, as we cannot even conclude $T > P$ if $S < P + \delta$ or, equivalently, $S - \delta < P$. Although their argument makes a lot of sense, the requirement to show $S > P + \delta$ raises the concern that δ has not been chosen properly, as δ should be no larger than the effect size $(S - P)$ (see Section 2.4 in Chapter 2) (EMEA/CPMP 2005; U.S. FDA 2010; and Ng 2008). To prespecify the NI margin, one has to resort to the historical data. This is necessary when planning the study for sample size calculation. Once the study is completed, there is no reason to use such a margin in the hypothesis testing, such as the fixed-margin approach in the ST design, because the margin may be incorporated into the hypotheses, as shown by Equations 8.2 and 8.3. Using the NI margin specified by Equation 2.1 in Section 2.2 of Chapter 2 implicitly assumes that $(S - P) > 0$ because the NI margin has to be positive. However, a mandatory requirement of showing $S > P$ is controversial, as discussed in Section 8.2.2.

For the ratio-based approach, however, one must be sure not to divide by zero in the ratio $(T–P)/(S - P)$, or in statistical terms, one must have enough confidence that $S > P$ (Röhmel 2005a, 2005b). On the other hand, the linearized approach does not have this limitation because the hypotheses defined by Equation 8.2 are expressed as a linear contrast of (T, S, and P) (Pigeot et al. 2003). In fact, the linearized approach compares T with the weighted average of S and P. Therefore, there is no requirement that $S > P$, although the concepts of NI and preservation do not make any sense if $S \leq P$. As long as the point estimate of $(S - P)$ is greater than zero, the requirement to show $S > P$ is less critical when the linearized approach is used than the ratio-based approach, because Fiellers's confidence interval used in the latter approach may be invalid when $(S - P)$ is close to zero. Therefore, the linearized approach is preferred.

8.3 Two Test Treatments versus an Active Control

The presentation in this section can easily be extended to comparisons of k tests versus an active control (or standard therapy, S). Furthermore, a one-sided version of the hypotheses may be formulated, but will not be discussed here.

Suppose that we are comparing two test treatments versus an active control. A natural way to do this is to compare each of the test groups with the control by testing the following two null hypotheses:

$$H_{10} : |T_1 - S| \geq \delta \text{ against } H_{11} : |T_1 - S| < \delta$$

and

$$H_{20} : |T_2 - S| \geq \delta \text{ against } H_{21} : |T_2 - S| < \delta$$

The first one is comparing the first test group versus the control, and the second one is comparing the second test group versus the control. Alternatively, these hypotheses can be combined as testing the null hypothesis

$$H_0 : \max_i |T_i - S| \geq \delta$$

against the alternative hypothesis

$$H_1 : \max_i |T_i - S| < \delta$$

Let X_1, X_2, and X_c denote the sample means of the first and second test treatments and the active control, respectively, and let n_1, n_2, and n_c denote the corresponding sample sizes. A simple way to test H_0 is to do pairwise comparisons and reject H_0 at the relevant significance level α. In other words, we would reject H_0 at a significance level of α if

$$(|X_1 - X_c| - \delta)/SE_1 < t(\alpha, n_1 + n_c - 2)$$

and

$$(|X_2 - X_c| - \delta)/SE_2 < t(\alpha, n_2 + n_c - 2)$$

where the standard errors SE_1 and SE_2 are computed by pooling pairs of sample variances. (Note that SE_1 is not the standard error of $|X_1 - X_c| - \delta$.) The standard errors can also be computed by pooling all three sample variances to achieve a larger degree of freedom. Assuming equal sample size and equal variance, Giani and Straßburger (1994) proposed to reject H_0 if

$$\max_i |X_i - X_c| < c_{GS}$$

where the critical value c_{GS} depends on the number of test groups and the significance level α. It also depends on δ, the sample size (n), and the variance (σ^2) through the expression $\delta n^{1/2}/\sigma$. Fiola and Wiens (1999) proposed a test statistic to allow for different sample sizes and/or different variances. They proposed to reject H_0 if

$$\max_i(|X_i - X_c| - \delta)/\sqrt{\sigma_i^2/n_i + \sigma_c^2/n_c} < c_{FW}$$

where the critical value c_{FW} is determined such that

$$\Pr\{\max_i(|X_i - X_c| - \delta)/[\min_i(\sigma_i^2/n_i) + \sigma_c^2/n_c]^{1/2} < c_{FW}\} \leq \alpha$$

Basically, the test statistic is defined as the maximum of the pairwise Z-statistics, and the critical value is calculated based on the minimum true standard deviation of pairwise sample mean differences. In these two papers, the critical value and/or the test statistic depend on the variances, which, in practice, have to be estimated.

8.4 Equivalence of Three Test Groups

8.4.1 Background/Introduction

Pairwise comparisons are often performed in studies involving three or more treatment groups. Such comparisons are sensible when two or more test products are compared with an active control, as discussed in Section 8.3. In lot consistency/release studies, however, there is no such control group for use in the pairwise comparisons. Even so, all possible pairwise comparisons can still be considered in such situations, resulting in a testing procedure based on a range statistic (see Section 8.4.3). Ng (2002) proposed an alternative approach based on the F-statistics from an analysis of variance (see Section 8.4.5). Since the critical value of the proposed test depends on the variance, Ng (2002) provided a method for finding the critical value iteratively when the variance must be estimated (see Section 8.4.6). See Lachenbruch, Rida, and Kou (2004) for a brief overview on lot consistency testing.

8.4.2 Assumptions, Notations, and Basic ANOVA Results

Suppose that we are comparing k ($k \geq 3$) treatment groups. We assume normality with a common variance σ^2. Let μ_i be the population mean for the ith treatment group ($i = 1,..., k$), and $\boldsymbol{\mu} = (\mu_1,..., \mu_k)'$ denotes the vector of the k population means. Suppose that there are k independent samples with equal sample size n. Let $\bar{X}_i$ be the sample mean for the ith treatment group ($i = 1,..., k$), let S^2 be the pooled sample variance, and let $\bar{X}$ be the grand sample mean.

The following are the basic results from one-way analysis of variance (ANOVA). We have

$$(n/\sigma^2)\,\Sigma_i\,(\bar{X}_i - \bar{X})^2 \sim \chi^2(k-1,\lambda) \tag{8.4}$$

and

$$[k(n-1)]\,S^2/\sigma^2 \sim \chi^2(k(n-1),0) \tag{8.5}$$

where $\chi^2(\nu,\lambda)$ denotes the noncentral chi-square distribution with noncentrality parameter λ and

$$\lambda = (n/\sigma^2)\,\Sigma_i\,(\mu_i - \bar{\mu})^2 \tag{8.6}$$

Furthermore, the two chi-square distributions are independent. We then form the F-statistic from these two chi-square distributions as follows:

$$F \equiv (n/S^2)\,\Sigma_i\,(\bar{x}_i - \bar{X})^2/(k-1) \tag{8.7}$$

This F-statistic follows a noncentral F-distribution with $(k-1)$ and $k(n-1)$ degrees of freedom and a noncentrality parameter λ given by Equation 8.6.

8.4.3 Pairwise-Based Methods

Now, we are comparing three test groups without a control. We can test all possible pairwise hypotheses in the following:

$$H_{10}:|\mu_1 - \mu_2| \geq \delta \text{ against } H_{11}:|\mu_1 - \mu_2| < \delta$$
$$H_{20}:|\mu_2 - \mu_3| \geq \delta \text{ against } H_{21}:|\mu_2 - \mu_3| < \delta$$

and

$$H_{30}:|\mu_3 - \mu_1| \geq \delta \text{ against } H_{31}:|\mu_3 - \mu_1| < \delta$$

We conclude that the three test groups are equivalent at a significance level of α if all the pairwise null hypotheses are rejected at α level; that is,

$$T_1 \equiv (|\bar{X}_1 - \bar{X}_2| - \delta)/SE_{12} < t(\alpha, n_1 + n_2 - 2)$$
$$T_2 \equiv (|\bar{X}_2 - \bar{X}_3| - \delta)/SE_{23} < t(\alpha, n_2 + n_3 - 2)$$

and

$$T_3 \equiv (|\bar{X}_3 - \bar{X}_1| - \delta)/SE_{31} < t(\alpha, n_3 + n_1\ 2)$$

where the standard errors SE_{12}, SE_{23}, and SE_{31} are computed by pooling pairs of sample variances. Again, we can pool all three sample variances to achieve larger degrees of freedom for estimating the variance. Alternatively, we can combine these pairwise hypotheses into the following:

$$H_0 : \max_{ij} |\mu_i - \mu_j| \geq \delta$$

against

$$H_1 : \max_{ij} |\mu_i - \mu_j| < \delta$$

or equivalently,

$$H_0 : \mu_{(3)} - \mu_{(1)} \geq \delta$$

against

$$H_1 : \mu_{(3)} - \mu_{(1)} < \delta$$

since $\max_{ij} |\mu_i - \mu_j| = \mu_{(3)} - \mu_{(1)}$, where $\mu_{(1)} < \mu_{(2)} < \mu_{(3)}$. Giani and Finner (1991) proposed the range statistic in a situation with an equal sample size and equal variance. The null hypothesis H_0 is rejected if

$$\bar{X}_{(3)} - \bar{X}_{(1)} < C_{GF}$$

The critical value c_{GF} is defined as $a\delta$, where "a" depends on the number of test groups and the significance level α. It also depends on the NI margin (δ), the sample size (n), and the variance (σ^2) through the expression $\delta n^{1/2}/\sigma$.

Wiens, Heyse, and Matthews (1996) modified the range statistic by subtracting δ and then dividing the result by the pairwise standard error. Wiens and Iglewicz (1999) proposed a test statistic to allow for a situation with an unequal sample size and/or unequal variance. They proposed to reject H_0 if

$$\max_{ij} (\bar{X}_i - \bar{X}_j - \delta)/[\sigma_i^2/n_i + \sigma_j^2/n_j]^{1/2} < c_{wI}$$

where

$$Pr\{(\bar{X}_{(3)} - \bar{X}_{(1)} - \delta)/[2\min_i(\sigma_i^2/n_i)]^{1/2} < c_{wI}\} \leq \alpha$$

Basically, the test statistic is defined as the maximum pairwise Z-statistics, and the critical value is calculated based on the minimum true standard deviation of pairwise sample mean differences. In these three papers, the critical value and/or the test statistic depend on the variances, which, in practice, have to be estimated.

8.4.4 Measures of Closeness

With two treatment groups, we can use the absolute main difference (i.e., $|\mu_t - \mu_c|$) to measure the closeness between the two treatments. How can the closeness be defined among three or more treatments? If we have defined the closeness for three means, denoted by $d(\mu_1, \mu_2, \mu_3)$, then we can test the null hypothesis

$$H_0 : d(\mu_1, \mu_2, \mu_3) \geq \delta$$

against the alternative hypothesis

$$H_1 : d(\mu_1, \mu_2, \mu_3) < \delta$$

In fact, $d(\mu_1, \mu_2, \mu_3) \equiv \mu_{(3)} - \mu_{(1)}$ is a measure of closeness among three treatments that was used by Giani and Finner (1991), as discussed in Section 8.4.3. An alternative measure of closeness among three treatments proposed by Ng (2000) will be discussed next.

First, let us consider the two-dimensional situation shown in Figure 8.1. For any point (μ_1, μ_2) within the two dotted lines, we have $|\mu_1 - \mu_2| < \delta$, whereas on the boundary, we have $|\mu_1 - \mu_2| = \delta$. However, δ is not the shortest distance from (μ_1, μ_2) to the 45-degree line passing through the origin. The shortest distance is given by

$$[(\mu_1 - \bar{\mu})^2 + (\mu_2 - \bar{\mu})^2]^{\frac{1}{2}}$$

which is the distance between (μ_1, μ_2) and $(\bar{\mu}, \bar{\mu})$, where $\bar{\mu} = (\mu_1 + \mu_2)/2$ and $(\bar{\mu}, \bar{\mu})$ is the projection of (μ_1, μ_2) onto the 45-degree line passing through the origin. This concept can be extended to three or more dimensions and will be discussed next.

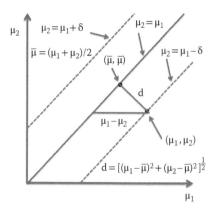

FIGURE 8.1
Measure of closeness (two treatment groups).

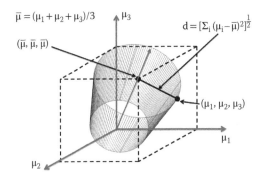

FIGURE 8.2
Measure of closeness (three treatment groups).

A measure of closeness among k means is defined as

$$d(\boldsymbol{\mu}) \equiv [\Sigma_i \, (\mu_i - \bar{\mu})^2]^{\frac{1}{2}}$$

which is the distance between $\boldsymbol{\mu}$ and $(\bar{\mu}, \dots, \bar{\mu}')$, where

$$\bar{\mu} = (\mu_1 + \dots + \mu_k)/k,$$

and $(\bar{\mu}_1, \dots, \bar{\mu})'$ is the projection of $\boldsymbol{\mu}$ onto the vector $(1, \dots, 1)'$. This is shown graphically in Figure 8.2 for $k = 3$. Setting $d(\boldsymbol{\mu}) = \delta$ would result in a circular cylinder with radius δ as shown in Figure 8.2. Within the cylinder, we have

$$d(\boldsymbol{\mu}) < \delta$$

and outside the cylinder, we have

$$d(\boldsymbol{\mu}) > \delta$$

8.4.5 Test for Equivalence of k Groups

We test the null hypothesis

$$H_0 : d(\boldsymbol{\mu}) \equiv [\Sigma_i \, (\mu_i - \bar{\mu})^2]^{\frac{1}{2}} \geq \delta \tag{8.8}$$

against the alternative hypothesis

$$H_1 : d(\boldsymbol{\mu}) \equiv [\Sigma_i \, (\mu_i - \bar{\mu})^2]^{\frac{1}{2}} < \delta$$

where δ (> 0) is prespecified. Ng (2000) proposed to reject H_0 at a significance level of α and conclude that $\boldsymbol{\mu}$ is within the cylinder if the F-statistic given by Equation 8.7 is less than the αth percentile of the noncentral F-distribution, with $(k-1)$ and $k(n-1)$ degrees of freedom and the noncentrality parameter given by

$$\lambda_0 = (n/\sigma^2)\Sigma_i\,(\mu_i - \bar{\mu})^2 = n\delta^2/\sigma^2 \qquad (8.9)$$

The second equality in Equation 8.9 follows under H_0 at the boundary. Since the null distribution for testing H_0 given by Equation 8.8 has a noncentral F-distribution, the critical value depends on σ^2, which must be estimated. Using an estimated critical value would inflate the Type I error rate. An iterative approach proposed by Ng (1993) is used to resolve this problem, which will be discussed next. Note that the null distribution for testing equality of k means in one-way ANOVA has a central F-distribution.

8.4.6 An Iterative Approach

The critical values and the Type I error rates are determined for $n = 20$, $k = 3$, $\delta = 1.5$, and $\alpha = 0.025$ to illustrate the iterative approach. Briefly, we start with a test statistic $T_1 \equiv \Sigma_i\,(\bar{X}_i - \bar{X})^2$ and determine the critical value as a function of σ^2. We then define a test statistic by subtracting the estimated critical value from T_1, and calculate its critical value as a function of σ^2. This iterative process stops when the critical value becomes reasonably flat and the inflation of the Type I error rate becomes negligible. The details of this iterative procedure are as follows:

In step 1, we start with

$$T_1 \equiv \Sigma_i\,(\bar{X}_i - \bar{X})^2$$

From Equation 8.4, the critical value of T_1, say, $c_1(\sigma^2)$, such that

$$P[T_1 \le c_1(\sigma^2)] = \alpha$$

can be determined as

$$c_1(\sigma^2) = (\sigma^2/n)\,\chi^{-1}(\alpha; k-1, \lambda_0) \qquad (8.10)$$

where $\chi^{-1}(\alpha; k-1, \lambda_0)$ denotes the αth percentile of the noncentral chi-square distribution, with $(k-1)$ degrees of freedom and a noncentrality parameter λ_0 given by Equation 8.9. The critical value $c_1(\sigma^2)$ is shown graphically in Figure 8.3a. However, when we use the estimated critical value for T_1, the Type I error rate will be inflated, as shown in Figure 8.3b. Thus, we proceed to step 2.

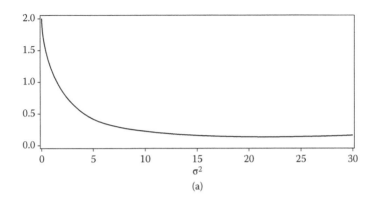

(a)

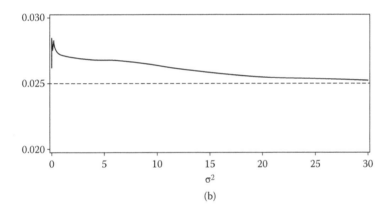

(b)

FIGURE 8.3
(a) Critical value: c_1. ($n = 20$, $k = 3$, $\delta = 1.5$, $\alpha = 0.025$). (b) Type 1 error rate using estimated c_1. ($n = 20$, $k = 3$, $\delta = 1.5$, $\alpha = 0.025$).

In step 2, we define

$$T_2 \equiv T_1 - c_1(S^2) \tag{8.11}$$

and then find the critical value of T_2, say, $c_2(\sigma^2)$, such that

$$P[T_2 \leq c_2(\sigma^2)] = \alpha$$

The computation of $c_2(\sigma^2)$ involves the distribution function of T_2, which will be derived in Appendix 8.A. The critical value $c_2(\sigma^2)$ is shown graphically in Figure 8.4a. When we use the estimated critical value for T_2, the Type I error rate will be either inflated or deflated, as shown in Figure 8.4b. Thus, we proceed to step 3.

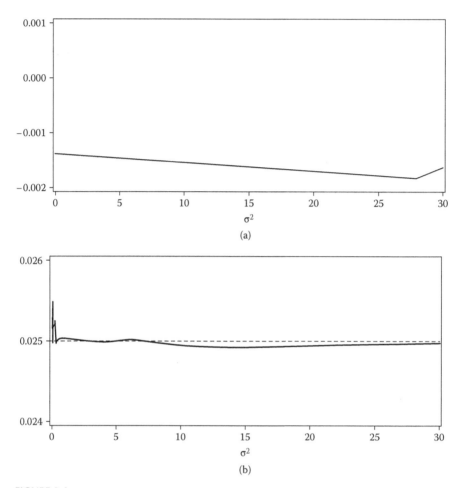

FIGURE 8.4
(a) Critical value: c_2. ($n = 20$, $k = 3$, $\delta = 1.5$, $\alpha = 0.025$). (b) Type 1 error rate using estimated c_2. ($n = 20$, $k = 3$, $\delta = 1.5$, $\alpha = 0.025$).

In step 3, we define

$$T_3 \equiv T_2 - c_2(S^2)$$

and then find the critical value of T_3, say, $c_3(\sigma^2)$, such that

$$P[T_3 \le c_3(\sigma^2)] = \alpha$$

The critical value $c_3(\sigma^2)$ is shown graphically in Figure 8.5a. When we use the estimated critical value for T_3, the Type I error rate will be either inflated or deflated, as shown in Figure 8.5b. Thus, we proceed to step 4. Note that T_3 can be expressed in terms of T_1 as follows:

$$T_3 \equiv T_2 - c_2(S^2) = T_1 - c_1(S^2) - c_2(S^2)$$

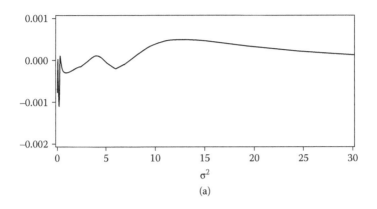

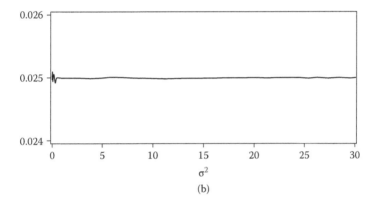

FIGURE 8.5
(a) Critical value: c_3. ($n = 20$, $k = 3$, $\delta = 1.5$, $\alpha = 0.025$). (b) Type 1 error rate using estimated c_3. ($n = 20$, $k = 3$, $\delta = 1.5$, $\alpha = 0.025$).

Continue this process. In step j, we define T_j as

$$T_j \equiv T_{j-1} - c_{j-1}(\sigma^2)$$

and then find the critical value of T_j, say, $c_j(\sigma^2)$, such that

$$P[T_j \leq c_j(\sigma^2)] = \alpha$$

When we use the estimated critical value for T_j, the Type I error rate will be either inflated or deflated. Thus, we proceed to step $(j + 1)$. Note that T_j can be expressed in terms of T_1 as follows:

$$T_j \equiv T_{j-1} - c_{j-1}(S^2)$$

$$= T_1 - c_1(S^2) - \ldots - c_{j-1}(S^2) \qquad (8.12)$$

The computation of $c_j(\sigma^2)$, for $j > 3$ involves the distribution function of T_j, which will be derived in Appendix 8.B. Hopefully, at some step, say, J,

$$c_J(\sigma^2) \cong 0$$

for all σ^2. We then stop and reject H_0 if

$$T_J \leq c_J(S^2)$$

The inflation of the Type I error rate will be negligible because

$$c_J(S^2) \cong c_J(\sigma^2)$$

for all σ^2. If we express T_J in terms of T_1 as given by Equation 8.12, we would reject H_0 if

$$T_1 < c^*(S^2)$$

where

$$c^*(\sigma^2) \equiv c_1(\sigma^2) + \ldots + c_j(\sigma^2)$$

8.4.7 Computation of $c_j(\sigma^2)$

Note that $c_1(\sigma^2)$ is given by Equation 8.10. Appendix 8.A gives a derivation of the distribution function of T_2 from which the critical value $c_2(\sigma^2)$ may be determined numerically. The distribution function of T_2 involves a double integral of one central and one noncentral chi-square distribution density function. Appendix 8.B gives a derivation of the distribution function of T_j ($j \geq 3$) from which the critical value $c_j(\sigma^2)$ may be determined numerically. The distribution function of T_j also involves a double integral of one central and one noncentral chi-square distribution density function. SAS/IML software is used in the computations of the critical values and the Type I error rates.

8.4.8 Discussion and Further Research

Based on the numerical experience for the one sample t-test and the Behrens-Fisher problem in Ng (1993), it is conjecture that $c_j(\sigma^2)$ would converge to zero uniformly as j approaches infinity because T_1 and S^2 are independent. This conjecture is also supported by the numerical results, as shown in Figures 8.6a and 8.6b.

With unequal sample sizes, it is not clear how to test the null hypothesis given by Equation 8.8 because the F-statistic for the one-way ANOVA with unequal sample sizes is given by

$$(1/S^2) \sum_i n_i (\bar{X}_i - \bar{X})^2 / (k-1)$$

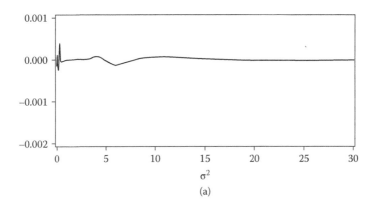

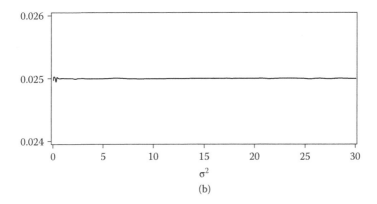

FIGURE 8.6

(a) Critical value: c_4. ($n = 20$, $k = 3$, $\delta = 1.5$, $\alpha = 0.025$). (b) Type 1 error rate using estimated c_4. ($n = 20$, $k = 3$, $\delta = 1.5$, $\alpha = 0.025$).

where n_i denotes the sample size of the ith treatment group. This statistic follows a noncentral F-distribution, with a noncentrality parameter given by

$$(1/\sigma^2)\sum_i n_i (\mu_i - \bar{\mu})^2$$

Therefore, the null distribution of the test statistic at the boundary depends on the μ_i's, which cannot be expressed as a function of δ. Nonetheless, using the geometric mean sample size may provide a good approximate solution when the sample sizes are not too unbalanced.

When comparing three or more dose regimens, δ may be determined by a small fraction of the overall expected effect size of the dose regimens. Furthermore, it is clear that a fixed-effect model—in the sense that the means are fixed as opposed to a random-effects model—is applicable in comparing dose regimens.

On the other hand, it is not clear how to choose δ for lot consistency studies. It may be more appropriate to assume a random-effects model. In that case, the main interest is to show that the future lots will not vary "too much" in terms of their true means. In other words, we would be testing the variability of the true means. This is an interesting research topic.

Appendix 8.A A Derivation of the Distribution Function of T_2

Let $h_2(t_2)$ be the distribution function of T_2 defined by Equation 8.11 under H_0 at the boundary. Write λ_0 given by Equation 8.9 as a function of σ^2 and δ as follows:

$$\lambda_0 \equiv \lambda_0(\sigma^2, \delta) = n\delta^2/\sigma^2$$

Using Equations 8.11, 8.10, and 8.4, we have

$$h_2(t_2) \equiv P[T_2 \leq t_2] = P[T_1 - c_1(S^2) \leq t_2] = P[T_1 - (S^2/n)\,\chi^{-1}(\alpha, k-1, \lambda_0(S^2, \delta)) \leq t_2]$$

$$= P[(n/\sigma^2)T_1 - (S^2/\sigma^2)\,\chi^{-1}(\alpha; k-1, \lambda_0(S^2, \delta)) \leq (n/\sigma^2)t_2]$$

$$= P[\chi^2(k-1, \lambda_0(\sigma^2, \delta)) \leq (n/\sigma^2)t_2 + (S^2/\sigma^2)\chi^{-1}(\alpha; k-1, \lambda_0(S^2, \delta))]$$

Expressing S^2 as a chi-square distribution in Equation 8.5, and due to the fact that the two chi-square distributions are independent, we have

$$h_2(t_2) = \int_0^\infty \int_0^{u(y;\,t_2)} f_x(x)\, f_y(y)\, dx\, dy$$

where $f_x(x)$ and $f_y(y)$ denote the probability density functions of $\chi^2(k-1, \lambda_0(\sigma^2, \delta))$ and $\chi^2(k(n-1), 0)$, respectively,

$$u(y; t_2) = (n/\sigma^2)\,t_2 + \chi^{-1}(\alpha; k-1, \lambda^*)y/[k(n-1)]$$

and

$$\lambda^* = k(n-1)n\delta^2/(y\sigma^2)$$

Appendix 8.B A Derivation of the Distribution Function of T_j ($j \geq 3$)

Let $h_j(t_j)$ be the distribution function of T_j defined by Equation 8.12 under H_0 at the boundary for $j \geq 3$. From Equations 8.12 and 8.4, we have

$$h_j(t_j) \equiv P[T_j \leq t_j] = P[T_1 - c_1(S^2) \ldots c_{j-1}(S^2) \leq t_j]$$

$$= P[(n/\sigma^2)T_1 - (n/\sigma^2)\{c_1(S^2) + \ldots + c_{j-1}(S^2)\} \leq (n/\sigma^2)t_j]$$

$$= P[\chi^2(k-1,\lambda_0(\sigma^2,\delta)) \leq (n/\sigma^2)t_j + (n/\sigma^2)\{c_1(S^2) + \ldots + c_{j-1}(S^2)\}]$$

Expressing S^2 as a chi-square distribution in Equation 8.5, and due to the fact that the two chi-square distributions are independent, we have

$$h_j(t_j) = \int_0^\infty \int_0^{u_j(y;\,t_j)} f_x(x)\, f_y(y)\, dx\, dy$$

where $f_x(x)$ and $f_y(y)$ are defined as in Appendix 8.A

$$u_j(y;\, t_j) = (n/\sigma^2)t_j + (n/\sigma^2)\{c_1(y^*) + \ldots + c_{j-1}(y^*)\}]$$

and

$$y^* = \sigma^2 y/[k(n-1)]$$

References

Committee for Proprietary Medicinal Products (CPMP; 2000). Points to Consider on Application with 1. Meta-Analyses; 2. One Pivotal Trial. CPMP/EWP/2330/99.

Committee for Proprietary Medicinal Products (CPMP; 2002a). Note for Guidance on Clinical Investigation of Medicinal Products in the Treatment of Depression. CPMP/EWP/518/97 Revision 1.

Committee for Proprietary Medicinal Products (CPMP; 2002b). Note for Guidance on the Clinical Investigation of Medicinal Products in the Treatment of Asthma. CPMP/EWP/2922/01. EMEA London.

European Agency for the Evaluation of Medicinal Products, Committee for Proprietary Medicinal Products (EMEA/CPMP, 2005). *Guideline on the Choice of the Non-Inferiority Margin EMEA/CPMP/EWP/2158/99.* http://www.ema .europa.eu/docs/en_GB/document_library/Scientific_guideline/2009/09 /WC500003636.pdf (Accessed: August 25, 2013).

Fiola MJ, and Wiens BL (1999). Testing Equivalence of Multiple Treatments with a Control. *Proceedings of the Biopharmaceutical Section, American Statistical Association*, 75–78.

Giani G, and Finner H (1991). Some General Results on Least Favorable Parameter Configurations with Special Reference to Equivalence Testing and the Range Statistic. *Journal of Statistical Planning and Inference*, **28**: 33–47.

Giani G, and Straßburger K (1994). Testing and Selecting for Equivalence with Respect to a Control. *Journal of the American Statistical Association*, **89**(425):320–329.

Hassalblad V, and Kong DF (2001). Statistical Methods for Comparison to Placebo in Active-Control Trials. *Drug Information Journal*, 35:435–449.

Hauschke D, and Pigeot I (2005). Establishing Efficacy of a New Experimental Treatment in the "Gold-Standard" Design. *Biometrical Journal*, 47:782–786.

Hida E, and Tango T (2011a). On the Three-Arm Non-inferiority Trial Including a Placebo with a Prespecified Margin. *Statistics in Medicine*, 30:224–231.

Hida E, and Tango T (2011b). Authors' Reply: Response to Joachim Röhmel and Iris Pigeot. *Statistics in Medicine*, 30:3165.

Hommel G (1988). A Stagewise Rejective Multiple Test Procedure Based on a Modified Bonferroni Test. *Biometrika*, **75**:383–386.

International Conference on Harmonization (ICH) E10 Guideline (2001). *Choice of Control Groups in Clinical Trials*. http://www.fda.gov/downloads/Drugs/GuidanceComplianceRegulatoryInformation/Guidances/UCM073139.pdf (Accessed: September 27, 2012).

Koch A (2005). Discussion on "Establishing Efficacy of a New Experimental Treatment in the 'Gold-Standard' Design." *Biometrical Journal*, 47:792–793.

Koch A, and Rohmel J (2004). Hypothesis Testing in the "Gold-Standard" Design for Proving the Efficacy of an Experimental Treatment Relative to Placebo and a Reference. *Journal of Biopharmaceutical Statistics*, **14**:315–325.

Lachenbruch PA, Rida W, and Kou J (2004). Lot Consistency as an Equivalence Problem, *Journal of Biopharmaceutical Statistics*, **14**:275–290.

Ng T-H (1993). A Solution to Hypothesis Testing Involving Nuisance Parameters. *Proceedings of the Biopharmaceutical Section, American Statistical Association*, 204–211.

Ng T-H (2000). Equivalence Testing with Three or More Treatment Groups. *Proceedings of the Biopharmaceutical Section, American Statistical Association*, 150–160.

Ng T-H (2001). Choice of Delta in Equivalence Testing. *Drug Information Journal*, **35**:1517–1527.

Ng T-H (2002). Iterative Chi-square Test for Equivalence of Multiple Treatment Groups, *Proceedings of the Biopharmaceutical Section, American Statistical Association*, 2464–2469.

Ng T-H (2008). Noninferiority Hypotheses and Choice of Noninferiority Margin. *Statistics in Medicine*, 27:5392–5406.

Pigeot I, Schafer J, Rohmel J, and Hauschke D (2003). Assessing Non-inferiority of a New Treatment in a Three-Arm Clinical Trial Including Placebo. *Statistics in Medicine*, 22:883–900.

Röhmel J (2005a). Discussion on "Establishing Efficacy of a New Experimental Treatment in the Gold-Standard Design." *Biometrical Journal*, 47:790–791.

Röhmel J (2005b). On Confidence Bounds for the Ratio of Net Differences in the "Gold-Standard" Design with Reference, Experimental, and Placebo Treatment. *Biometrical Journal*, **47**:799–806.

Röhmel J, and Pigeot I (2010). A Comparison of Multiple Testing Procedures for the Gold-Standard Non-inferiority Trial. *Journal of Biopharmaceutical Statistics,* **20**:911–926.

Rohmel J, and Pigeot I (2011). Statistical Strategies for the Analysis of Clinical Trials with an Experimental Treatment, an Active Control and Placebo, and a Prespecified Fixed Non-inferiority Margin for the Difference in Means. Letter to the Editor. *Statistics in Medicine,* **30**:3162–3164.

Rothmann MD and Tsou HH (2003). On Non-inferiority Analysis Based on Delta Method Confidence Intervals. *Journal of Biopharmaceutical Statistics,* **13**:565–583.

Shaffer JP (1986). Modified Sequentially Rejective Multiple Test Procedures. *Journal of American Statistical Association,* **81**:826–831.

Temple R, and Ellenberg SS (2000). Placebo-Controlled Trials and Active-Control Trials in the Evaluation of New Treatments. *Annals of Internal Medicine,* **133**:464–470.

U.S. Food and Drug Administration (2010). *Draft Guidance for Industry: Non-inferiority Clinical Trials.* http://www.fda.gov/downloads/Drugs/GuidanceCompliance RegulatoryInformation/Guidances/UCM202140.pdf (Accessed: August 25, 2013).

Wiens B, Heyse J, and Matthews H (1996). Similarity of Three Treatments, with Application to Vaccine Development. *Proceedings of the Biopharmaceutical Section, American Statistical Association,* 203–206.

Wiens B, and Iglewicz B (1999). On Testing Equivalence of Three Populations. *Journal of Biopharmaceutical Statistics,* **9**(3):465–483.

9

Regulatory Guidances

9.1 Introduction

In the 1990s, two noninferiority (NI) guideline documents were released by regulatory agencies: one from U.S. Food and Drug Administration (FDA) (1992) and the other one from the European Agency for the Evaluation of Medicinal Products, Committee for Proprietary Medicinal Products (EMEA/CPMP) (1997). These two documents give advice on the NI margin in the anti-infective area with a binary endpoint. Since then, numerous guidelines regarding equivalence trials or NI trials have been released. Treadwell (2011) reviewed and summarized 12 publically available guidelines/documents (mostly from regulatory agencies) for individual trials that describe themselves as either equivalence trials or NI trials:

- Two documents from the International Conference on Harmonization (ICH): ICH E9 (1998) and ICH E10 (2001)
- Four documents from the European Medicines Agency (EMEA): EMEA/CPMP (2000a, 2000b, 2003, 2005)
- One document from each of the following regulatory agencies:
  a. Japan Pharmaceuticals and Medical Devices Agency (Japan PMDA 2008)
  b. The Australian Pharmaceutical Benefits Advisory Committee (Australian PBAC 2008)
  c. The U.S. Food and Drug Administration (U.S. FDA 2010)
  d. The India Central Drugs Standard Control Organisation (India CDSCO 2010)
- Two documents from collaborative academic groups: Gomberg-Maitland, Frison, and Halperin (2003) and Piaggio et al. (2006)

This undertaking is part of research conducted by the Evidence-based Practice Center (EPC) Workgroup (Treadwell, Uhl, Tipton, et al. 2012) with the objective to provide guidance on how to manage the concepts of equivalence and NI in the context of systematic reviews. One of the EMEA/CPMP documents specifically deals with the choice of NI margin (2005), while the U.S. FDA (2010) document provides guidance specifically for NI trials.

In this chapter, four regulatory documents will be discussed: U.S. FDA (1992), EMEA/CPMP (1997), EMEA/CPMP (2005), and U.S. FDA (2010). The focus is on determining the NI margin. The first two documents will be discussed in Section 9.2, and the last two documents will be discussed in Sections 9.3 and 9.4, respectively. Additional guideline documents are discussed in Section 9.2.3, including two FDA Anti-Infective Drugs Advisory Committee meetings held in 2002 and 2008.

9.2 Guidelines in the Anti-Infective Area with a Binary Endpoint

9.2.1 FDA and CPMP Guidelines

In 1992, the Division of Anti-Infective Drug Products of the FDA issued a points-to-consider document for the development and labeling of antimicrobials (U.S. FDA 1992 pp. 20–21). It stated the following:

> For primary clinical or microbiologic effectiveness endpoints with values greater than 90 percent for the better of the two drugs, a confidence interval that crosses zero and remains within the lower bound delta of −0.10 or less will usually be required to establish equivalence. For primary clinical or microbiologic effectiveness endpoints with values of 80 percent to 89 percent for the better of the two drugs, a confidence interval that crosses zero and remains within the lower bound delta of −0.15 or less will usually be required to establish equivalence. For primary clinical or microbiologic effectiveness endpoints with values of 70 percent to 79 percent for the better of the two drugs, a confidence interval that crosses zero and remains within the lower bound delta of −0.20 or less will usually be required to establish equivalence.

Since the 1990s, the vast majority of comparative trials submitted to the FDA used this so-called "step function" for selecting deltas (Powers et al. 2002).

In 1997, the CPMP issued a similar guideline (EMEA/CPMP 1997) for the evaluation of antibacterial substances. This guideline proposed an NI margin of 0.1 throughout, but also indicated that a smaller δ may be necessary for very high (>90 percent) response rates. It did not, however, specify how much smaller δ should be (Rohmel 1998, 2001).

When testing the NI hypothesis given in by Equation 1.3a in Section 1.7 of Chapter 1, both guidelines recommended using the lower limit of the 95% confidence interval (CI) for the treatment difference (T − S) in comparison to −δ. More specifically, the null hypothesis is rejected if the lower limit exceeds −δ. The FDA guideline also required the upper limit of the 95% CI to exceed 0, which was not the case for the CPMP guideline.

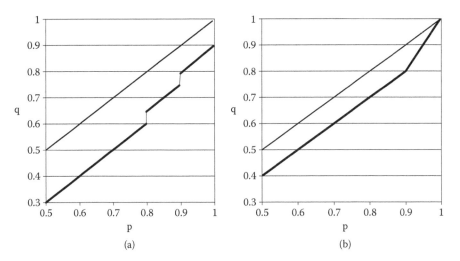

FIGURE 9.1
(a) FDA margin. (b) Adapted from the CPMP margin. (From Rohmel J. 2001. *Statistics in Medicine*; 20:2561–2571.)

Rohmel (2001) contrasted the two guidelines in terms of the margin in Figure 1 of his paper. This figure is reproduced here as Figure 9.1, where q and p correspond to T and S, respectively. Rohmel extended the range of p to 0.5 in Figure 9.1b for the FDA margin and arbitrarily connected the points (p = 0.9, q = 0.8) with (1.0, 1.0) by a straight line in Figure 9.1b for the CPMP margin. The thick lines in Figure 9.1 are the boundary of a subspace in the lower-right corner of the full parameter space. This is called the null space H_0. The only point for which both margins (FDA and CPMP) coincide is p = 0.9, q = 0.8, where $\delta = 0.1$. For all other points, the CPMP margin is smaller than the FDA margin.

Note that no explicit NI margin was recommended in a more recent EMEA/CPMP guideline (EMEA/CPMP 2011) or one issued by the U.S. FDA (see Section 9.2.3).

9.2.2 Varying NI Margin with S

As noted in Section 9.2.1, the CPMP guideline proposed an NI margin of 0.1 throughout, but also indicates that a smaller δ may be necessary for very high (>90 percent) response rates. The reason for a smaller δ when the response rates are high is apparently due to the concern of doubling the failure rate (i.e., T = 2S). See, for example, Section 4.2 in Chapter 4 and the following statements excerpted from DiNubile et al. (2012 pp. 1055–1056):

> For binary outcomes, when success rates approach 90%, [NI] margins of approximately 10% no longer seem appropriate as they imply that failure rates twice as high as that of the standard are acceptable.

...in circumstances where the standard of care reliably produces response rates of approximately 95%, [an NI] margin as small as 5% accepts a possible doubling of the failure rate with the new therapy.

There are two drawbacks with the FDA margin: (1) the margin as a function of S is not smooth, and (2) the margin is random, as it depends on the observed S (as well as the observed T). These lead to (1) an erratic behavior of the power curve (Rohmel 1998), and (2) a serious inflation of the Type I error rate (Rohmel 2001). Rohmel (2001) and Phillips (2003) proposed a smooth function of S as the NI margin. However, these proposals do not take the effect size (i.e., S − P) into consideration as suggested in Section 2.2 in Chapter 2 and by ICH E10 (2001) and the U.S. FDA (2010). One alternative to having a smooth varying NI margin is by formulating and testing the NI hypotheses using the odds ratio as the metric, as given by Equation 4.1 in Section 4.3 of Chapter 4.

9.2.3 Additional FDA Guidance Documents and Advisory Committee Meetings

In consideration of the ICH E10 guideline, in February 2001, the U.S. FDA stated on its website that the step-function method is no longer in use and that detailed guidelines on specifying deltas is under development (Powers et al. 2002). The revised points-to-consider document (U.S. FDA 2001, pp. 20–21) stated the following:

> The sliding scale method for determination of delta previously included in this document is no longer in use. Further general information on specifying delta can be found in the 1998 Draft Guidance on Developing Antimicrobial Drugs – General Considerations for Clinical Trials (Section XX.A.6) as well as in the 2000 ICH E10 document (section 1.5.1.1 entitled "Historical Evidence of Sensitivity to Drug Effects and Choosing the Noninferiority Margin").

> Detailed guidance about delta specification is currently under development. Sponsors are strongly encouraged to specify a delta in their protocols and provide a rationale for that choice. Consultation with the Review division is recommended.

The 1998 Draft Guidance (U.S. FDA 1998) emphasized using a "two-tailed 95% confidence interval around the difference in outcomes" approach to determine equivalence between two products. It stated in Section XX.A.6 ("Issues in Similarity, or *Equivalence* Trial Design") that "The lack of a *statistically significant* difference should not be used as evidence of similarity." However, there is no guideline on how to determine the delta.

In February 2002, the FDA held an Anti-Infective Drugs Advisory Committee meeting on an approach for selecting delta in NI (equivalence) clinical trials (U.S. FDA 2002). Many consensuses were reached in the meeting (Shlaes 2002), such as the following:

- Infectious disease indications are different and it is impractical to have a common statistical requirement across such indications.
- The issue of "biocreep" could be addressed by (1) assuring the use of appropriate comparator agents and (2) in certain cases in which a placebo effect might be strong, the use of a three-arm trial that uses a placebo control with early withdrawal.

It is unlikely that the historical data will support the determination of an NI margin for the indications of (1) acute bacterial sinusitis, (2) acute bacterial exacerbation of chronic bronchitis, and (3) acute bacterial otitis media (U.S. FDA 2010).

In November 2008, the FDA held another Anti-Infective Drugs Advisory Committee meeting on the justifications of the NI margin for complicated skin and skin structure infections (SSSI) (U.S. FDA 2008). The committee recommended that NI trials are acceptable in complicated SSSI. However, the committee recommended against using NI trials in uncomplicated SSSI.

9.3 EMEA/CPMP Document

The 2005 EMEA/CPMP document entitled "Guideline on the Choice of the Non-Inferiority Margin" is only 11 pages long, including one cover page and one page for the table of contents, leaving only 9 pages for the actual contents. The document describes five situations where an NI trial might be performed as opposed to, or in addition to, a superiority trial over placebo:

1. Applications based upon essential similarity in areas where bioequivalence studies are not possible (e.g., modified release products or topical preparations)
2. Products with a potential safety advantage over the standard might require an efficacy comparison to the standard to allow a risk-benefit assessment to be made
3. Cases where a direct comparison against the active comparator is needed to help make a risk-benefit assessment
4. Cases where no important loss of efficacy compared to the active comparator would be acceptable
5. Disease areas where the use of a placebo arm is not possible and an active-control trial is used to demonstrate the efficacy of the test product

In situations 2 and 3, the objective of the NI trial is to evaluate the efficacy of T relative to S to assess the risk-benefit ratio. Such an objective is discussed in Section 4 of the document and corresponds to Objective 2 stated in Section 8.2.1 of Chapter 8. Situation 5 is in disease areas where the use of a placebo arm is not possible (presumably for ethical reasons), and an active-control trial is used to demonstrate the efficacy of the test product by indirect comparison. This corresponds to Objective 1 stated in Section 8.2.1 of Chapter 8. These two objectives are also discussed in Section 2.4 of Chapter 2.

The document emphasizes that the aim of the trial should be precisely defined so that the NI margin may be chosen accordingly. This is in line with the proposed choice of NI margin in Sections 2.2 and 2.4 of Chapter 2, where the choice of ε depends on the study objective. The document states that trials are generally labeled NI trials if they are not aiming to show superiority over the reference. As methodologies showing indirect comparison of T with P emerge, the document points out that (1) "demonstrating [NI]" is not considered to be a sufficiently detailed objective for a trial, and (2) the conclusions of the trial should not be that "[NI]" has been demonstrated, but some more precise statement reflecting the objectives of the trial. Perhaps NI trials with Objective 1 should be labeled "efficacy NI trials."

Regarding specifying the NI margin as a small percentage of the effect size (see Section 2.2 of Chapter 2), the 2005 EMEA/CPMP document states the following:

- It is not appropriate to define the [NI] margin as a proportion of the difference between the active comparator and placebo. Such ideas were formulated with the aim of ensuring that the test product was superior to (a putative) placebo; however, they may not achieve this purpose. If the reference product has a large advantage over placebo, this does not mean that large differences are unimportant; it just means that the reference product is very efficacious (page 5).

- Alternatively, the aim may be to provide data to show that there is no important loss of efficacy if the test product is used instead of the reference. This is probably the most common aim of [NI] trials. The choice of delta for such an objective cannot be obtained by only looking at past trials of the comparator against placebo. Ideas such as choosing delta to be a percentage of the expected difference between active and placebo have been advocated, but this is not considered an acceptable justification for the choice. Such ideas were principally formulated to ensure that the reference product was superior to placebo, but this has already been addressed in section III of this document. To adequately choose delta, an informed decision must be taken, supported by evidence of what is considered an unimportant difference in the particular disease area (page 8).

- The main point is that the aim of the trial should be precisely defined. Following that, a choice for delta should be made, supported by evidence, based upon the precise objectives. This evidence will not solely come from past trials of the comparator against placebo. Of course, the final choice must always be at least as small as the value derived from the considerations of section III (page 9).

On one hand, the document appears to recognize the role of the effect size (i.e., S – P) in determining the NI margin. On the other hand, it is against using a fraction of the effect size as the NI margin, apparently due to the concern that the margin may be too large when the effect size is very large, even though the concept paper on choice of delta (EMEA/CPMP 1999) suggested a delta of one-half or one-third the effect size in some situations (see Item 10 in Section 1.4 of Chapter 1).

The document provides general considerations when choosing the NI margin. However, no explicit guidance on how to determine the NI margin is given. This is in contrast with the FDA draft guidance to be discussed next, which gives explicit guidance.

9.4 FDA Draft Guidance

The FDA draft guidance entitled "Guidance for Industry: Non-Inferiority Clinical Trials" was released for public comments in March 2010 and has yet to be finalized. It is 63 pages long, excluding the cover page and the table of contents, so it is seven times as long as the EMEA/CPMP guideline. The FDA document provides explicit guidance on how to set the NI margin, including four examples. It recommends first determining the value for M_1, defined as "treatment effect of the active comparator" (i.e., how well the active-control treatment is expected to work), and then calculating the value for M_2, referred to as the clinical margin, by taking a fraction of M_1.

Using "S" instead of "C" as the active control, the null hypothesis is stated as

$$H_0: S-T \geq M \tag{9.1}$$

versus the alternative hypothesis

$$H_a: S-T < M$$

where M is the NI margin that corresponds to δ and is set to M_1 or M_2. The null hypothesis given by Equation 9.1 is rejected if the upper limit of the 95% CI for (S – T) is less than M, or equivalently, the lower limit of the 95% CI for (T – S) is greater than –M.

Although it is not explicitly stated, it appears that $M_1 = S - P$. Taking a fraction of M_1 as M_2 is in line with the NI margin given by Equation 2.1 in Chapter 2. On the other hand, M_1 looks more like an estimate of S – P. In fact, M_1 is typically determined as the lower limit of the 95% confidence for the effect size in the historical trial, with discounting if needed

(see Section 2.5.2 of Chapter 2). For example, if 10% discounting is used
(i.e., $\gamma = 0.9$), then

$$M_1 = 0.9 \times \text{lower limit of the 95\% CI for } (S-P)_h$$

where $(S-P)_h$ denotes the effect size in the historical studies under the fixed-effect model (FEM) (see Section 7.3 of Chapter 7) or the global mean under the random-effects model (REM) (see Section 7.4 of Chapter 7), although this expression for M_1 is not explicitly stated in the draft guidance. A typical value for M_2 is often 50% of M_1 (i.e., $\varepsilon = 0.5$), at least partly because the sample sizes needed to rule out a smaller loss become impractically large (U.S. FDA 2010).

Although M_1 and M_2 depend on the historical data, they are considered fixed in Equation 9.1. Therefore, the hypothesis testing involving M_1 and M_2 is called the FEM (see Section 5.3 of Chapter 5). Without formulating the hypotheses involving the parameters of the historical data as in Section 5.4 of Chapter 5, the description of the synthesis method in the draft guidance is not clear. To contrast the differences between the two methods, the draft guidance refers to Example 1(B).

The draft guidance recommends the fixed-margin approach for ensuring that the test drug has an effect greater than placebo (i.e., the NI margin M_1 is ruled out), but recommends the synthesis approach for ruling out the clinical margin M_2. Strictly speaking, since M_2 is fixed, one cannot rule out a fixed margin such as M_2 using the synthesis approach.

References

Australia Pharmaceutical Benefits Advisory Committee (2008). Guidelines for Preparing Submissions to the Pharmaceutical Benefits Advisory Committee (Version 4.3) Barton ACT: Commonwealth of Australia; Dec. 2008. p. 300. http:// www.pbs.gov.au/industry/listing/elements/pbac-guidelines/PBAC4.3.2.pdf (Accessed: August 25, 2013).

DiNubile MJ, Sklar P, Lupinacci RJ, and Eron JJ Jr (2012). Paradoxical Interpretations of Noninferiority Studies: Violating the Excluded Middle. *Future Virology,* 7(11), 1055–1063.

European Agency for the Evaluation of Medicinal Products, Committee for Proprietary Medicinal Products (1997). Evaluation of New Anti-Bacterial Products.

European Agency for the Evaluation of Medicinal Products, Committee for Proprietary Medicinal Products (1999). Concept Paper on Points to Consider: Choice of Delta. CPMP/EWP/2158/99. http://www.f-mri.org/upload /module-5/STAT_CHMP2158_Delta_choice.pdf (Accessed: September 8, 2013).

European Agency for the Evaluation of Medicinal Products, Committee for Proprietary Medicinal Products (2000a). Points to Consider on Switching Between Superiority and Noninferiority. http://www.ema.europa.eu/docs /en_GB/document_library/Scientific_guideline/2009/09/WC500003658.pdf (Accessed: August 25, 2013).

European Agency for the Evaluation of Medicinal Products, Committee for Proprietary Medicinal Products (2000b). Note for Guidance on the Investigation of Bioavailability and Bioequivalence. London (U.K.); December 14, 2000. http://www.ema.europa.eu/docs/en_GB/document_library/Scientific _guideline/2009/09/WC500003519.pdf.

European Agency for the Evaluation of Medicinal Products, Committee for Proprietary Medicinal Products (2003). Common Technical Document for the Registration of Pharmaceuticals for Human Use: Clinical Overview and Clinical Summary of Module 2. Module 5: Study Reports. London (UK): July 2003, 44. http://www.ema.europa.eu/docs/en_GB/document_library/Scientific _guideline/2009/09/WC500002723.pdf (Accessed: August 25, 2013).

European Agency for the Evaluation of Medicinal Products, Committee for Proprietary Medicinal Products (2005). Guideline on the Choice of the Non-Inferiority Margin EMEA/CPMP/EWP/2158/99. http://www.ema.europa.eu/docs /en_GB/document_library/Scientific_guideline/2009/09/WC500003636.pdf (Accessed: August 25, 2013).

European Agency for the Evaluation of Medicinal Products, Committee for Proprietary Medicinal Products (2011). Guideline on the Evaluation of Medicinal Products Indicated for Treatment of Bacterial Infections. http://www.ema.europa.eu /docs/en_GB/document_library/Scientific_guideline/2009/09/WC500003417 .pdf (Accessed: April 30, 2014).

Gomberg-Maitland M, Frison L, and Halperin JL (2003). Active-control Clinical Trials to Establish Equivalence or Noninferiority: Methodological and Statistical Concepts Linked to Quality. *American Heart Journal;* **146**(3):398–403.

India Central Drugs Standard Control Organization (2010). Guidance for Industry on Preparation of Common Technical Document for Import/Manufacture and Marketing Approval of New Drugs for Human Use (New Drug Application – NDA), New Delhi; November 2010, 110. http://cdsco.nic.in/CTD _Guidance%20-Final.pdf (Accessed: August 25, 2013).

International Conference on Harmonization (ICH) E9 Guideline (1998). *Statistical Principles for Clinical Trials.* http://www.fda.gov/downloads/Drugs /GuidanceComplianceRegulatoryInformation/Guidances/UCM073137.pdf (Accessed: September 27, 2012).

International Conference on Harmonization (ICH) E10 Guideline (2001). *Choice of Control Groups in Clinical Trials.* http://www.fda.gov/downloads/Drugs /GuidanceComplianceRegulatoryInformation/Guidances/UCM073139.pdf (Accessed: September 27, 2012).

Japan Pharmaceuticals and Medical Devices Agency (2008). Points to Be Considered by the Review Staff Involved in the Evaluation Process of New Drugs. Final. April 17, 2008. http://www.pmda.go.jp/english/service/pdf/review/points .pdf (Accessed: May 1, 2014).

Phillips KF (2003). A New Test of Non-inferiority for Anti-infective Trials. *Statistics in Medicine;* **22**:201–212.

Piaggio G, Elbourne DR, Altman DG, Pocock SJ, and Evans SJ (2006). Reporting of Noninferiority and Equivalence Randomized Trials: An Extension of the CONSORT Statement. *Journal of American Medical Association,* **295**:1152–1160.

Powers JH, Ross DB, Brittain E, Albrecht R, and Goldberger MJ (2002). The United States Food and Drug Administration and Noninferiority Margins in Clinical Trials of Antimicrobial Agents. *Clinical Infectious Diseases;* **34**:879–881.

Rohmel J (1998). Therapeutic Equivalence Investigations: Statistical Considerations. *Statistics in Medicine,* **17**.1703 1714.

Rohmel J (2001). Statistical Considerations of FDA and CPMP Rules for the Investigation of New Anti-bacterial Products. *Statistics in Medicine;* **20**:2561–2571.

Shlaes DM (2002). Reply to comments by David N. Gilbert and John E. Edwards, Jr. on the paper: Shlaes DM, Moellering RC, Jr. The United States Food and Drug Administration and the end of antibiotics [letter]. Clin Infect Dis 2002; 34: 420–2. *Clinical Infectious Diseases;* **35**:216–217.

Treadwell J, Uhl S, Tipton K, Singh S, Santaguida L, Sun X, Berkman N, Viswanathan M, Coleman C, Shamliyan T, Wang S, Ramakrishnan R, and Elshaug A (2012). Assessing Equivalence and Noninferiority. Methods Research Report. (Prepared by the EPC Workgroup under Contract No. 290-2007-10063.) AHRQ Publication No. 12-EHC045-EF. Rockville, MD: Agency for Healthcare Research and Quality, June 2012. http://www.effectivehealthcare.ahrq.gov/ehc/products/365/1154 /Assessing-Equivalence-and-Noninferiority_FinalReport_20120613.pdf (Accessed: March 1, 2014).

Treadwell JR (2011). Methods Project 1: Existing Guidance for Individual Trials. Plymouth Meeting, PA: ECRI Institute; June 29, 2011, 39.

U.S. Food and Drug Administration (1992), Division of Anti-infective Drug Products. Points-to-Consider: Clinical Development and Labeling of Anti-Infective Drug Products (Rescinded February 2001). http://www.fda.gov/ohrms/dockets /ac/02/briefing/3837b1_09_1992%20version.pdf (Accessed: September 2, 2013).

U.S. Food and Drug Administration (1998). Draft Guidance for Industry: Developing Antimicrobial Drugs – General Considerations of Clinical Trials. July 1998. http://www.fda.gov/downloads/Drugs/GuidanceComplianceRegulatory Information/=Guidances/ucm070983.pdf (Accessed: September 4, 2013).

U.S. Food and Drug Administration (2001), Division of Anti-infective Drug Products. Points-to-Consider: Clinical Development and Labeling of Anti-Infective Drug Products. Disclaimer of 1992 Points-to-Consider Document. http://www.fda.gov /ohrms/dockets/ac/02/briefing/3837b1_11_points%20to%20consider%20 GFI.pdf (Accessed: September 2, 2013).

U.S. Food and Drug Administration (2002), Division of Anti-Infective Drug Products Advisory Committee Meeting, February 19–20, 2002. http://www.fda.gov /ohrms/dockets/ac/cder02.htm#Anti-Infective (Accessed: September 2, 2013).

U.S. Food and Drug Administration (2010). Guidance for Industry – Antibacterial Drug Products: Use of Non-inferiority Studies to Support Approval. http://www .fda.gov/downloads/Drugs/GuidanceComplianceRegulatoryInformation/ Guidances/ucm070951.pdf (Accessed: August 16, 2014).

U.S. Food and Drug Administration (2008), Division of Anti-Infective Drug Products Advisory Committee Meeting, November 19–20, 2008. (http://www.fda.gov /ohrms/dockets/ac/cder08.html#AntiInfective (Accessed: September 2, 2013).

U.S. Food and Drug Administration (2010). Draft Guidance for Industry: Non-inferiority Clinical Trials. http://www.fda.gov/downloads/Drugs /GuidanceComplianceRegulatoryInformation/Guidances/UCM202140.pdf (Accessed: August 25, 2013).

10

Intention-to-Treat versus Per-Protocol

10.1 Introduction

As stated by Lewis and Machin (1993, p. 647), "In the clinical researcher's perfect world, every subject entered into a randomized controlled clinical trial (RCT) would satisfy all entry criteria, would complete their allocated treatment as described in the protocol, and would contribute data records that were complete in all respects." However, since clinical trials are dealing with human subjects, strict adherence to the trial protocol is impossible. Nonadherence to the trial protocol can be broadly classified into four categories related to four aspects of the trial: (1) inclusion/exclusion criteria, (2) study medication/treatment, (3) scheduled visit, and (4) efficacy measurement. These are explained further as follows:

1. Subjects are found to be ineligible (i.e., do not meet the inclusion/exclusion criteria) after randomization. One possible reason is due to misdiagnosis of the patient condition.

2. Subjects do not take study medication as scheduled or do not receive the specified amount of the treatment/intervention (noncompliance). At the extreme, subjects (1) do not take any study medication or receive any treatment/intervention, (2) receive the alternative treatment, or (3) take prohibited concomitant medications. Possible reasons for noncompliance could be toxicity, dropouts, or withdrawal.

3. Subjects miss the visits or do not make the visits as scheduled. Subjects may withdraw or drop out due to death, relocation, or loss to follow-up.

4. No efficacy measurements are taken, primarily due to missed visits.

In clinical trials, there are different forms of medical intervention and different study endpoints, even within the same therapeutic area, let alone across different therapeutic areas. The most common intervention is drug products in the form of pill for many different diseases/sicknesses. Other interventions include (1) surgical procedures, which are particularly common in cancer research; (2) transfusion with biological products (e.g., blood and blood-related products); (3) medical devices, etc. The study endpoints include (1) cure of the disease or resolution of illness; (2) progression-free or long-term survival, which is particularly common in cancer research; (3) quality of life, etc. For a given RCT, depending upon the therapeutic areas (and/or different forms of medical intervention and different study endpoints), each of the four categories of nonadherence may or may not be applicable.

Data analysis in RCTs is complex, challenging, and controversial due to non-adherence to trial protocol, and it is evolving. The complexity arises because various aspects of nonadherence (see earlier) need to be addressed, and each therapeutic area (or each form of medical intervention and/or different study endpoints within a therapeutic area) may have its own unique issues. The concepts of (1) efficacy versus effectiveness (see Section 10.2.2) and (2) the explanatory approach versus the pragmatic approach (see Section 10.2.3) add another layer of complexity. Nonadherence to trial protocol leads to two major issues in data analysis: missing data and noncompliance. The controversy is due to the intention-to-treat (ITT) principle (see Section 10.2.1). There are two major aspects of data analysis: (1) subjects are analyzed as "randomized" (see Sections 10.3.2 and 10.6.1) or as "treated" (see Section 10.3.4), and (2) some randomized subjects may or may not be excluded from analysis.

The ITT principle (see Section 10.2.1) was controversial and prompted considerable debate for decades (see Section 10.4.1). Following the ITT principle, all randomized subjects were included in an analysis as "randomized" (see Section 10.3.2), while a per-protocol (PP) analysis excluded protocol violators (see Section 10.3.3). The controversy about whether to use an ITT or a PP approach in superiority trials largely subsided in the 1990s in favor of support for ITT (see Section 10.4.1). On the other hand, there is no clear "winner" for NI trials (see Section 10.5). Current thinking from regulatory agencies is that both analyses are of equal importance and should lead to similar conclusions for a robust interpretation (see Section 10.5.1). However, using primarily ITT analysis in NI trials is gaining support in recent publications (e.g., Fleming et al. 2011; Schumi and Wittes 2011) (see Section 10.5.5). Implementation of an ITT analysis in RCTs is discussed in Section 10.6.

10.2 Background

10.2.1 Intention-to-Treat Principle

The phrase "intention to treat" (ITT) was originated by Sir Austin Bradford Hill, and the term "intent-to-treat" is commonly used (Rothmann, Wiens, and Chan 2011). Bradford-Hill recommended that all participants be included "in the comparison and thus measure the intention to treat in a given way rather than the actual treatment" (1961, 258) as he noted that post-randomization exclusion of participants could affect the internal validity that randomization sought to achieve. As stated by Polit and Gillespie (2010, p. 357), "His advice was primarily aimed at researchers who deliberately removed subjects from the analysis."

Cochrane Collaboration (2002, p. 5) reiterated the ITT principle as follows:

> The basic intention-to-treat principle is that participants in trials should be analyzed in the groups to which they were randomized, regardless of whether they received or adhered to the allocated intervention.

A glossary in ICH E9 (1998, p. 42) included "intention-to-treat principle" as follows:

> The principle asserts that the effect of a treatment policy can be best assessed by evaluating on the basis of the intention to treat a subject (i.e., the planned treatment regimen) rather than the actual treatment given. It has the consequence that subjects allocated to a treatment group should be followed up, assessed, and analyzed as members of that group irrespective of their compliance with the planned course of treatment.

Gillings and Koch (1991, p. 411) elaborated the ITT principle as follows:

> The fundamental idea behind ITT is that exclusion from the statistical analysis of some patients who were randomized to treatment may induce bias which favors one treatment group more than another. Bias may also occur if patients are not analyzed as though they belonged to the treatment group originally intended by the randomization procedure rather than the treatment actually received.

Deviations from the ITT principle, such as (1) excluding randomized subjects, as in PP analysis (see Section 10.3.3), and/or (2) transferring subjects from one group to another group, as in as-treated analysis (see Section 10.3.4), would destroy the comparability of the treatment group that is achieved through randomization (Newell 1992).

10.2.2 Efficacy versus Effectiveness

Roland and Torgerson (1998, p. 285) define (1) efficacy as the "benefit a treatment produces under ideal conditions, often using carefully defined subjects in a research clinic," and (2) effectiveness as "the benefit the treatment produces in routine clinical practice." Hernan and Hernandez-Diaz (2012, p. 50) define efficacy as "how well a treatment works under perfect adherence and highly controlled conditions," and effectiveness as "how well a treatment works in everyday practice." Additionally, "Effectiveness takes into consideration how easy a drug is to use, and potential side effects, whereas efficacy measures only how well it produces the desired result" (Neal, 2009). Thaul (2012, p. 4) elaborates the differences between efficacy and effectiveness in the following:

> *Efficacy* refers to whether a drug demonstrates a health benefit over a placebo or other intervention when tested in an ideal situation, such as a tightly

controlled clinical trial. *Effectiveness* describes how the drug works in a real-world situation. Effectiveness is often lower than efficacy because of interactions with other medications or health conditions of the patient, sufficient dose or duration of use not prescribed by the physician or followed by the patient, or use for an off-label condition that had not been tested.

Roland and Torgerson (1998) introduce two different types of clinical trials: (1) explanatory trials, which measure efficacy, and (2) pragmatic trials, which measure effectiveness. Section 10.2.3 discusses the different aspects of these two types of trials.

10.2.3 Explanatory Approach versus Pragmatic Approach

Roland and Torgerson (1998, p. 285) describe differences between the explanatory and pragmatic approaches in different aspects of the trials, such as patient population, study endpoint and analysis set, as follows:

- Patient population. An explanatory approach recruits as homogeneous a population as possible and aims primarily to further scientific knowledge. By contrast, the design of a pragmatic trial reflects variations between patients that occur in real clinical practice and aims to inform choices between treatments. To ensure generalizability, pragmatic trials should, so far as possible, represent the patients to whom the treatment will be applied. The need for purchasers and providers of health care to use evidence from trials in policy decisions has increased the focus on pragmatic trials.

- Study endpoint. In explanatory trials, intermediate outcomes are often used, which may relate to understanding the biological basis of the response to the treatment—for example, a reduction in blood pressure. In pragmatic trials, outcomes should represent the full range of health gains—for example, a reduction in stroke and improvement in quality of life.

- Analysis set. In a pragmatic trial, it is neither necessary nor always desirable for all subjects to complete the trial in the group to which they were allocated. However, patients are always analyzed in the group to which they were initially randomized (intention-to-treat analysis), even if they drop out of the study or change groups.

Schwartz and Lellouch (1967) describe the differences in study design between the two approaches and "consider a trial of anticancer treatments in which radiotherapy alone is to be compared with radiotherapy preceded by the administration of a drug that has no effect by itself but that may sensitize the patient to the effects of radiation." The drug is assumed to be administered over a 30-day period. The "radiotherapy alone" group may then be handled in two different ways (see Figure 10.1), as described by Schwartz and Lellouch (1967, p. 638):

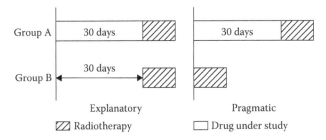

FIGURE 10.1
Explanatory and pragmatic approaches. (From Schwartz D, and Lellouch J. 1967. *Journal of Chronic Diseases*, **20**:637–648.)

1. Radiotherapy may be preceded by a blank period of 30 days so that it is instituted at the same time in each group.
2. Radiotherapy may be instituted at once, thereby carrying it out at what is most probably the optimal time.

In design 1 (delayed radiotherapy in both groups), the two groups are alike from the radiotherapy point of view and differ solely in the presence or absence of the drug. Therefore, it provides an assessment of the sensitizing effect of the drug and gives valuable information at a biological level. This is the explanatory approach. It provides an answer to the research question whether the drug has a sensitizing effect. However, it does not provide an answer to a practical question whether the combined treatment is better than immediate radiotherapy.

In design 2 (immediate radiotherapy in one group), the two treatments are compared under the conditions in which they would be applied in practice. This is the pragmatic approach. It provides an answer to a practical question whether the combined treatment is better than immediate radiotherapy. However, it will provide information on the effectiveness of the drug only when the combined treatment proves to be better than radiotherapy alone.

10.3 Analyses of Randomized Control Trials

10.3.1 A Simplified Schema for a Randomized Control Trial

Referring to Figure 10.2, Newell (1992, pp. 838–839) described a simplified schema for a randomized control trial as follows:

> For any RCT of a health care intervention (e.g., formal rehabilitation compared with none, day-case surgery compared with inpatient care, two different recruitment methods for mammography), the broad research outline is as shown in [Figure 10.2]... by the end of the trial, there were four groups of patients: (1) those allocated to A who did not complete A, (2) those allocated to A who did complete it, (3) those allocated to B who completed it, and (4) those

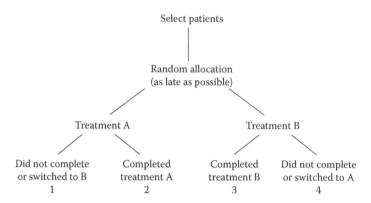

FIGURE 10.2
A simplified schema for a randomized controlled trial. (From Newell, D. J. 1992. *International Journal of Epidemiology*, **21**:837–841.)

allocated to B who did not complete it. ITT analysis (otherwise known as "pragmatic trial" or "[program] effectiveness analysis") compares 1 + 2 with 3 + 4. Efficacy analysis (otherwise known as "explanatory trial" or "test of biological efficacy") compares 2 with 3, ignoring 1 and 4. Treatment-received analysis (otherwise known as "as treated") compares 1 + 3 with 2 + 4 when treatments are switched.

These three types of analyses are further elaborated in the next three subsections.

10.3.2 Intention-to-Treat Analyses

An ITT analysis is also known as a program effectiveness analysis (see Sections 10.2.2 and 10.3.1). It includes all randomized subjects in the treatment groups to which they were randomized, regardless of whether they received or adhered to the assigned treatment. Such an analysis preserves comparable treatment groups due to randomization and prevents bias resulting from post-randomization exclusions. Newell (1992, p. 837) provides the rationale for the ITT analysis in a different way, as follows:

> The purpose of randomization is to avoid selection bias and to generate groups which are comparable to each other. Any changes to these groups by removing some individuals' records or transferring them to another group destroy that comparability.

Comparability of treatment groups is the foundation on which the statistical inference is built. Validity of the statistical inference will be compromised if such comparability is destroyed.

In theory, there is a consensus in the literature regarding the definition of an ITT analysis, as illustrated in the following:

- "In 1990, a work group for the Biopharmaceutical Section of the American Statistical Association (ASA) came to the conclusion that it is one which *includes all randomized patients in the groups to which they were randomly assigned, regardless of their compliance with the entry criteria, regardless of the treatment they actually received, and regardless of subsequent withdrawal from treatment or deviation from the protocol* (Fisher et al. 1990)" (Lewis and Machin 1993, p. 647).

- "In an intent-to-treat analysis (ITT), patients are analyzed according to the treatment to which they were assigned, regardless of whether they received the assigned treatment" (D'Agostino, Massaro, and Sullivan 2003, p. 182).

- "The ITT analysis includes all randomized patients in the groups to which they were randomly assigned, regardless of their compliance with the entry criteria, the treatment they actually received, and subsequent withdrawal from treatment or deviation from the protocol" (Le Henanff et al. 2006, p. 1148).

- "In an ITT analysis, subjects are analyzed according to their assigned treatment regardless of whether they actually complied with the treatment regimen" (Sheng and Kim 2006, p. 1148).

- "Intention-to-treat (ITT) is an approach to the analysis of randomized controlled trials (RCT) in which patients are analyzed as randomized regardless of the treatment actually received" (Gravel, Opartny, and Shapiro, 2007, p. 350).

- "In the ITT approach, all patients we intended to treat will be included into the analysis, whether they completed the trial following the protocol or not" (Gonzalez, Bolaños, and de Sereday 2009).

- "According to the principle, trial participants should be analyzed within the study group to which they were originally allocated irrespective of non-compliance or deviations from protocol" (Alshurafa et al. 2012).

- "An ITT analysis includes all participants according to the treatment to which they have been randomized, even if they do not receive the treatment. Protocol violators, patients who miss one or more visits, patients who drop out, and patients who were randomized into the wrong group are analyzed according to the planned treatment" (Treadwell et al. 2012, pp. 7–8).

In practice, however, a strict ITT analysis is often hard to achieve for two main reasons: missing outcomes for some participants and nonadherence to the trial protocol (Moher et al. 2010). Different definitions of modified ITT analysis emerge due to nonadherence to the trial protocol by excluding some randomized patients, as given in the following:

- "…In a further 25 RCTs in 1997–1998, a modified intent-to-treat analysis was performed, which excluded participants who never received treatment or who were never evaluated while receiving treatment" (Hill, LaValley, and Felson 2002, p. 783).

- "Some definitions of ITT exclude patients who never received treatment" (D'Agostino, Massaro, and Sullivan 2003, p. 182).

- "A *true or classic* ITT is one that removes none of the subjects from the final analysis— with the exception of ineligible subjects removed post-randomization…" (Polit and Gillespie 2010, p. 357).

- "In this article, modified ITT refers to an approach in which all participants are included in the groups to which they were randomized, and the researchers make efforts to obtain outcome data for all participants, even if they did not complete the full intervention (Gravel et al. 2007; Polit and Gillespie 2009)" (Polit and Gillespie 2010, pp. 357–358).

To include all randomized patients in an ITT analysis, imputation methods are often used to deal with missing data. However, such methods require an untestable assumption of "missing at random" to some degree. The best way to deal with the problem is to have as little missing data as possible (Lachin 2000; NRC 2010). Although the ITT has been widely used as the primary analysis in superiority trials, it is often inadequately described and inadequately applied (Hollis and Campbell 1999). For example, studies that claimed use of ITT did not indicate how missing outcomes or deviations from protocols were handled.

10.3.3 Per-Protocol Analyses

In contrast to an ITT analysis, a per-protocol (PP) analysis excludes protocol violators and includes only participants who adhere to the protocol as defined by some authors in the following:

- "One analysis which is often contrasted with the ITT analysis is the 'per protocol' analysis. Such an analysis includes *only those patients who satisfy the entry criteria of the trial and who adhere to the protocol subsequently* (here again there is ample room for different interpretations of what constitutes adherence to the protocol)" (Lewis and Machin 1993, p. 648).

- "The per-protocol analysis includes only patients who satisfied the entry criteria of the trial and who completed the treatment as defined in the protocol" (Le Henanff et al. 2006, p. 1184).

- "In a PP approach, however, only subjects who completely adhered to the treatment are included in the analysis" (Sheng and Kim 2006, p. 1184).
- "… is preferred to per-protocol (PP) analysis (i.e., using outcomes from only those participants who fully complied with the study protocol)" (Scott 2009, p. 329).
- "… the per-protocol (PP) population, which in this case is the set of people who have taken their assigned treatment and adhered to it" (Schumi and Wittes 2011).

On the other hand, other authors exclude major protocol violators from the PP analyses and allow minor protocol violators, as defined in the following:

- "The per-protocol (PP) analysis includes all patients who completed the full course of assigned treatment and who had no major protocol violations" (D'Agostino, Massaro, and Sullivan 2003, p. 182).
- "The PP population is defined as a subset of the ITT population who completed the study without any major protocol violations" (Sanchez and Chen 2006, p. 1171).
- "In general, the PP analysis, as described by ICH E-9 guidance, includes all subjects who were, in retrospect, eligible for enrollment in the study without major protocol violations, who received an acceptable amount of test treatment, and who had some minimal amount of follow-up" (Wiens and Zhao 2007, p. 287).

Excluding major protocol violators and allowing minor protocol violations opens the door for subjective judgments as to (1) what constitutes minor or major protocol violations, (2) what constitutes an acceptable amount of test treatment, and (3) what constitutes a minimal amount of follow-up.

Presumably, subjects who were switched to the other treatment arm (or who were wrongly randomized) are considered major protocol violators. Therefore, such subjects would be excluded from the PP analysis, although it is not explicitly stated so in the aforementioned definitions, with the exception that is defined by Newell (1992) in Section 10.3.1 where the term "efficacy analysis" is used instead of "PP analysis."

PP (or efficacy) analysis is also known as test of biological efficacy (see Sections 10.2.2 and 10.3.1) or on-treatment analysis (Heritier, Gebski, and Keech 2003; Piaggio et al. 2006; Kaul and Diamond 2007).

10.3.4 As-Treated Analyses

One key feature in ITT analysis is that subjects are analyzed according to their assigned treatment (i.e., as-randomized). This is in contrast with "as-treated" (AT) analysis where subjects are analyzed according to the treatment received, regardless of the regimen to which they were assigned, including subjects who do not complete the trial and those who switch from one treatment to another (Wertz 1995; Wiens and Zhao 2007). It is also known as a "treatment-received analysis" and a "garbage analysis" (Newell 1992; Wertz 1995).

10.3.5 Analysis Sets

From a practical point of view, a strict ITT analysis, including all random-ized subjects, may not be warranted. In fact, ICH E9 (1998, p. 41) defines full analysis set as "the set of subjects that is as close as possible to the ideal implied by the intention-to-treat principle. It is derived from the set of all randomized subjects by minimal and justified elimination of subjects." The document presents three circumstances that might lead to excluding ran-domized subjects from the full analysis set: (1) the failure to satisfy major entry criteria (eligibility violations), (2) the failure to take at least one dose of trial medication, and (3) the lack of any data post-randomization, and states that such exclusions should always be justified. Peto et al. (1976) allow some inappropriately randomized patients to be excluded (see Section 10.6).

ICH E9 (1998, p. 41) defines per-protocol set (sometimes described as the valid cases, the efficacy sample, or the evaluable subjects sample) as

> the set of data generated by the subset of subjects who complied with the protocol sufficiently to ensure that these data would be likely to exhibit the effects of treatment according to the underlying scientific model. Compliance covers such considerations as exposure to treatment, avail-ability of measurements, and absence of major protocol violations.

10.4 Intention-to-Treat Analysis versus Per-Protocol Analysis in Superiority Trials

10.4.1 Controversy of the ITT Principle

In late 1980s, ITT analysis (see Section 10.3.2) was endorsed by regulatory agencies (U.S. FDA 1988; NCM 1989) as the primary analysis of RCT data. It is also recommended by the American Statistical Associations Group (Fisher et al. 1990) and the Cochrane Collaboration (Moher, Schulz, and Altman 2001). Note that such recommendation is in the context of placebo-controlled trials as opposed to active-control trials (more specifically, NI trials).

The goal of the ITT principle in RCTs is to preserve the prognostic balance between participants in treatment and control groups achieved through randomization and to thereby minimize selection bias and confounding (Alshurafa et al. 2012) (see also Sections 10.2.1 and 10.3.2). However, the ITT principle was controversial and prompted considerable debate for decades.

See, for example, Lachin (2000) and the references therein. The controversy is summarized by Polit and Gillespie (2010, p. 357) as follows:

> Sir Bradford-Hill's recommendation was controversial and instigated considerable debate (Lachin 2000). Opponents advocated removing subjects who did not receive the treatment, arguing that such a per-protocol analysis would test the true efficacy of the intervention. Opponents maintained that it is not sensible to include in the intervention group people who did not actually receive the intervention. This position was most sharply expressed within the context of pharmaceutical trials, where estimates of effectiveness could be construed as the degree of beneficial effects among those who were compliant and able to tolerate the drug. Those advocating an ITT analysis, on the other hand, insisted that per-protocol analyses would likely lead to biased estimates of effectiveness because removal of noncompliant patients undermined the balance that randomization was presumed to have created (the methodological argument). Moreover, they argued that an ITT approach yields more realistic estimates of average treatment effects, inasmuch as patients in the real world drop out of treatment or fail to comply with a regimen (the clinical and policy argument).

One argument for ITT analysis in dealing with noncompliance in superiority trials is that noncompliance would lower the apparent impact of effective interventions and hence, would provide a conservative estimate of the treatment effect (Alshurafa et al. 2012). Although such an argument in superiority trials is legitimate, it is problematic in NI trials because a conservative estimate of the treatment effect may actually increase the likelihood of falsely concluding noninferiority, that is, inflation of the Type I error rate (see Section 10.5). Another argument for an ITT analysis, independent of study design (superiority or NI), is that the purpose of the analysis is to estimate the effects of allocating an intervention in practice, not the effects in the subgroup of participants who adhere to it (Cochrane Collaboration 2002). In other words, the interest is to assess effectiveness rather than efficacy (see Section 10.2.2).

10.4.2 Recommendations for ITT

As stated by Lewis and Machin (1993, p. 648), "... ITT is better regarded as a complete trial strategy for design, conduct, and analysis rather than as an approach to analysis alone." In this regard, Hollis and Campbell (1999, p. 674) recommend the following:

- Design
  - Decide whether the aim is pragmatic or explanatory. For pragmatic trials, ITT is essential.
  - Justify in advance any inclusion criteria that, when violated, would merit exclusion from ITT analysis.

- Conduct
  - Minimize missing response on the primary outcome.
  - Follow up with subjects who withdraw from treatment.

- Analysis
  - Include all randomized subjects in the groups to which they were allocated.
  - Investigate the potential effect of missing response.

- Reporting
  - Specify that ITT analysis has been carried out, explicitly describing the handling of deviations from randomized allocation and missing response.
  - Report deviations from randomized allocation and missing response.
  - Discuss the potential effect of missing response.
  - Base conclusions on the results of ITT analysis.

See Schulz, Altman, and Moher (2010) for the CONSORT guidelines for reporting parallel-group randomized trials.

10.4.3 Discussion

There are two strong arguments in favor of ITT analysis over PP analysis of superiority trials: (1) the ITT analysis preserves randomization, while exclusion of noncompliant subjects in the PP analysis could introduce bias that randomization intends to avoid; and (2) the ITT analysis including noncompliant subjects is more reflective of "real-world" practice, while the PP analysis is not, as it excludes noncompliant subjects. With regard to the latter argument, the ITT analysis including noncompliant subjects will result in conservative estimates of efficacy, which is not the case in the estimation of effectiveness by definition (see Section 10.2.2).

A strict ITT analysis includes all randomized subjects with the "once randomized always analyzed" philosophy (Schulz and Grimes 2002) as "randomized," while a strict PP analysis includes only subjects who strictly adhere to the study protocol. Therefore, these two analyses are at the two extremes in terms of the total sample size, which is not necessarily true in terms of bias. Assessing bias in estimating the efficacy due to noncompliance is complex and is beyond the scope of this book. More importantly, from regulatory and practical points of view, adjustment for noncompliance is not warranted, as the interest is in effectiveness rather than efficacy (see Section 10.2.2). Readers are referred to an issue of *Statistics in Medicine* for the analysis of compliance (volume 17, number 3, 1998).

Various modified ITT analyses have been used in practice (see Section 10.3.2) that allow some randomized subjects to be excluded. On the other hand, some PP analyses include minor protocol violators (see Section 10.3.3). For example, Gillings and Koch (1991) define (1) the ITT population as all randomized patients who were known to take at least one dose of treatment and who provided any follow-up data for one or more key efficacy variables, and (2) the efficacy analyzable population as a subset of the ITT population who adhered to the critical aspects of the protocol. Strict adherence to all protocol features is needed for a second version of the efficacy analyzable population. The authors recommend that patients be analyzed according to the treatments actually received when only a few patients are wrongly randomized and such administrative errors are not associated with the background characteristics of these patients or their prognosis.

10.5 Intention-to-Treat Analysis versus Per-Protocol Analysis in Noninferiority Trials

10.5.1 Current Thinking from Regulatory Agencies

Although ITT analysis is widely accepted as the primary analysis in superiority trials (see Sections 10.1 and 10.4), there is concern that inclusion of noncompliant subjects in such an analysis in NI (or equivalence) trials would dilute the potential treatment difference, leading to an erroneous conclusion of NI, and therefore, it is anticonservative (see Section 10.5.2). On the other hand, inclusion of only compliant subjects (i.e., exclusion of noncompliant subjects) in a PP analysis would reflect the treatment differences that may exist; however, such exclusion could undermine the prognostic balance between the two treatment arms achieved through randomization, leading to potential biases (see Section 10.3.2). Thus, for NI trials, there is no single ideal analysis strategy in the face of substantial noncompliance or missing data, and both analysis by ITT and well-defined PP analyses would seem warranted (Pocock 2003).

The European Agency for the Evaluation of Medicinal Products, Committee for Proprietary Medicinal Products (EMEA/CPMP) guidance (2000) states that in an NI trial, the full analysis set and the PP analysis set have equal importance and their use should lead to similar conclusions for a robust interpretation. It should be noted that the need to exclude a substantial proportion of the ITT population from the PP analysis throws some doubt on the overall validity of the study (CPMP 1995). The Food and Drug Administration (FDA) draft guidance (2010) states the following, noting that "as-treated" analysis is not defined in the document (see Section 10.3.4 for this definition):

> In [NI] trials, many kinds of problems fatal to a superiority trial, such as
> nonadherence, misclassification of the primary endpoint, or measurement

problems more generally (i.e., "noise"), or many dropouts who must be assessed as part of the treated group, can bias toward no treatment difference (success) and undermine the validity of the trial, creating apparent [NI] where it did not really exist. Although an "as-treated" analysis is therefore often suggested as the primary analysis for NI studies, there are also significant concerns with the possibility of informative censoring in an as-treated analysis. It is therefore important to conduct both ITT and as-treated analyses in NI studies.

10.5.2 Anticonservatism of ITT Analysis in Noninferiority Trials

Although ITT analysis is widely accepted as the primary analysis in superiority trials, as discussed in Section 10.4, it was recognized in the 1990s that ITT analysis plays a different role in NI (or equivalence) trials because it is anticonservative in such cases. For example, Lewis and Machin (1993) stated the following:

> ... These are often called equivalence trials. ... In such a trial, an ITT analysis generally increases the chance of erroneously concluding that no difference exists. When we are comparing an active agent with placebo, this increased risk is acceptable and is deliberately incurred. In these trials, ITT is conservative; we only declare a new agent effective when we have incontrovertible evidence that this is so, and the inevitable dilution of the treatment effect in an ITT analysis makes it harder to achieve this goal and affords extra statistical protection for the cautious. But when we are seeking equivalence, the bias is in an anticonservative direction.

And ICH E9 (1998) stated the following with some edits in the first sentence:

> The full analysis set and the per-protocol set [see Section 10.3.5] play different roles in superiority trials and in equivalence or [NI] trials. In superiority trials, the full analysis set is used in the primary analysis (apart from exceptional circumstances) because it tends to avoid overoptimistic estimates of efficacy resulting from a per-protocol analysis. This is because the noncompliers included in the full analysis set will generally diminish the estimated treatment effect. However, in an equivalence or [NI] trial, use of the full analysis set is generally not conservative and its role should be considered very carefully.

The ITT analysis could be anticonservative in poorly conducted NI trials. For example, mixing up treatment assignments would bias toward similarity of the two treatments when the test treatment is not effective or is less effective than the active control (Ng 2001) (see Sections 1.5.3 in Chapter 1 and 2.5.1 in Chapter 2). Schumi and Wittes (2011) elaborated on such a hypothetical example in the following:

> Consider a trial with a hopelessly flawed randomization, where instead of creating two distinct treatment groups (one set of subjects receiving

the new treatment and the other the active comparator), the randomization scheme actually created two "blended" groups, each composed of half [the] subjects receiving the new treatment and half receiving the active comparator. If this trial were testing for superiority, the test would, with high probability, correctly find no difference between the groups. As [an NI] trial, however, such a flawed trial would be very likely to incorrectly demonstrate [NI].

Lewis and Machin (1993) gave an extreme hypothetical example where all patients were withdrawn from both randomized arms and put on the same standard therapy; an ITT analysis would conclude that no difference existed between the original treatments, regardless of their true relative efficacy. In another extreme hypothetical example given by Brittain and Lin (2005) where no subject complies with therapy, NI could be "demonstrated" between any two therapies with an ITT analysis. In fact, there is consensus regarding the role of ITT in NI trials, as seen by many authors:

1. "In comparative studies [which seek to show one drug to be superior] the ITT analysis usually tends to avoid the optimistic estimate of efficacy which may result from a PP analysis, since the noncompliers included in an ITT analysis will generally diminish the overall treatment effect. However, in an equivalence trial, ITT no longer provides a conservative strategy and its role should be considered very carefully" (CPMP 1995, p. 1674).

2. "For equivalence trials, however, there is concern that an ITT analysis will move the estimated treatment difference towards zero since it will include patients who should not have been in the trial who will get no benefit, or patients who did not get the true treatment benefit because of protocol violation or failure to complete. PP analyses include only those who follow the protocol adequately. This would be expected to detect a clearer effect of treatment since uninformative 'noise' would be removed" (Ebbutt and Frith 1998, p. 1699).

3. "In [NI] trials, the ITT analysis tends to be 'liberal.' That is, by inclusion of those who do not complete the full course of the treatments, the ITT tends to bias towards making the two treatments (new treatment and active control) look similar. The PP removes these patients and is more likely to reflect differences between the two treatments" (D'Agostino, Massaro, and Sullivan 2003, p. 182).

4. "The dilemma for [NI] trials is that faced with non-negligible non-compliance analysis by [ITT] could artificially enhance the claim of [NI] by diluting some real treatment difference" (Pocock 2003, p. 489).

5. "[ITT] analyses will generally be biased toward finding no difference, which is usually the desired outcome in [NI] and equivalence trials and is favored by studies with many dropouts and missing data" (Gøtzsche 2006, p. 1173).

6. "However, in the [NI] setting, because the null and alternative hypotheses are reversed, a dilution of the treatment effect actually favors the alternative hypothesis, making it more likely that true inferiority is masked. An alternative approach is to use a [PP] population, defined as only participants who comply with the protocol" (Greene et al. 2008, p. 473).

7. "For [an NI] trial, the ITT analysis does not have a conservative effect. Dropouts and a poor conduct of the study might direct the results of the two arms toward each other. Another possibility is to consider the PP population, which consists of only the nonprotocol violators" (Lesaffre 2008, p. 154).

8. "In an [NI] trial, ITT analysis is thus more likely to narrow the difference between treatments and yield a noninferior result. Consequently, a PP analysis is needed to cross-validate the ITT analysis, while bearing in mind substantial variation between treatment groups in rates and reasons for dropout may also invalidate PP analyses" (Scott 2009, p. 329).

9. "Use of an ITT analysis in these trials could lead to a false conclusion of EQ-NI by diluting any real treatment differences" (Treadwell et al. 2012, p. 8).

10.5.3 Role of Per-Protocol Analyses in Noninferiority Trials

Since an ITT analysis could be anticonservative in poorly conducted NI trials, as discussed in Section 10.5.2, PP analysis provides an alternative that would reflect the treatment difference that may exist by excluding noncompliant subjects. See, for example, items 2, 3, 6, 7, and 8 in Section 10.5.2. For this reason, use of the PP population as the primary analysis in NI trials became prominent in the early 2000s (Garrett 2003).

Many authors (e.g., Ebbutt and Frith 1998; Gøtzsche 2006) recognize that excluding noncompliant subjects in the PP analysis could undermine the prognostic balance between the two treatment arms achieved through randomization, leading to potential biases. This concern is the key reason why ITT analysis rather than PP analysis is widely accepted as the primary analysis for superiority trials (see Section 10.4.1). This leads to the regulatory agencies and many authors (e.g., Lewis and Machin 1993; Pocock 2003; Pater 2004; Kaul and Diamond 2007; Lesaffre 2008; and Scott, 2009) recommending that both ITT and PP analyses be performed in NI trials (see Section 10.5.1). For example, Lewis and Machin (1993, p. 649) stated the following:

> However, careful consideration of the alternatives does not lead to totally abandoning the ITT analysis. Indeed, in equivalence trials, the overall ITT strategy of collecting all endpoints in all randomized patients is equally valuable, but the role of the ITT analysis itself differs. Essentially, the ITT

analysis and one or more plausible [PP] analyses need to provide the same conclusion of no difference before a safe overall conclusion can be drawn.

10.5.4 Examples of Comparisons between Intent-to-Treat and Per-Protocol Analyses in Noninferiority and Equivalence Trials

Ebbutt and Frith (1998) compared the ITT and PP results in 11 asthma equivalence trials where peak expiratory flow rate was used as the efficacy endpoint. The PP analysis consistently gave wider confidence intervals than the ITT analysis due entirely to a smaller number of subjects included in the PP analysis. There was no evidence that the ITT analyses were more conservative in their estimates of treatment difference. The authors suggested that the relative importance of the two analyses will depend on the definitions used in particular therapeutic areas and recommended seeking prior agreement with regulatory agencies on the role of ITT and PP populations.

Brittain and Lin (2005) compared the ITT and PP analyses from 20 antibiotic trials that were presented at the FDA Anti-Infective Drug Products Advisory Committee from October 1999 through January 2003. They saw no indication that the PP analysis tends to produce a larger absolute treatment effect than the ITT analysis; on the contrary, the data hints at the possibility that there may be a tendency for the ITT analysis to produce a larger observed treatment effect than the PP analysis. They speculated that in typical studies of antibiotic therapy, both analyses may often underestimate the "pure" treatment difference; that is, the effect in the set of patients who comply with therapy and have no other complicating factors. Note that "pure" treatment difference refers to efficacy as opposed to effectiveness (see Section 10.2.2).

In a systematic review, Wangge et al. (2010) included 227 articles on NI trials registered in PubMed on February 5, 2009. These articles referred to 232 trials. In 97 (41.8%) of the trials, both ITT and PP analyses were performed. They did not observe any evidence that the ITT analysis will lead to more NI conclusions than the PP analysis and concluded that both analyses are equally important, as each approach brings a different interpretation for the drug in daily practice.

10.5.5 Intention-to-Treat as the Primary Analysis in Noninferiority Trials

Due to concerns with ITT and PP analyses in NI trials (see Sections 10.5.2 and 10.5.3, respectively), current thinking from regulatory agencies does not put more emphasis on one over the other (see Section 10.5.1). However, there is a need for one primary analysis in NI trials from the industry point of view, as expressed by some authors in the following:

- "The need to demonstrate equivalence in both an ITT and a PP analysis in a regulatory trial increases the regulatory burden on drug developers. The relative importance of the two analyses will depend on the definitions used in particular therapeutic areas. Demonstrating equivalence in one population with strong support from the other would be preferred from the industry viewpoint" (Ebbutt and Frith 1998, p. 1691).

- "The suggestion to use both ITT and PP analyses to evaluate active-control studies places extra burdens on sponsors to meet regulatory objectives. Making one analysis primary is preferable for efficient drug development and consistent regulatory decision making" (Wiens and Zhao 2007, p. 291).

Wiens and Zhao (2007) advocated the use of ITT as the primary analysis in NI trials. They noted that the reasons for using ITT as the primary analysis in superiority trials, such as (1) preserving the value of randomization and (2) estimating real-world effectiveness (see Sections 10.2.2 and 10.4.1) are also applicable in NI trials. Other reasons are as follows:

- In order to support the constancy assumption, the ITT analysis should be preferred for the NI study because the active-control effect used in the NI margin determination is likely to be estimated based on an ITT analysis.
- Inference in switching from NI to superiority will be much simpler if the same analysis set is used.
- A major change in philosophy for NI testing with a very small NI margin is not rational (Hauck and Anderson 1999; Wiens and Zhao 2007).

On the other hand, the authors also discussed reasons for not using ITT analyses and considered three general areas in which such analyses are potentially biased: enrolling ineligible subjects, poor study conduct, and missing data. They elaborated on how to address/mitigate these potential biases. Fleming (2008, p. 328) favored the ITT analysis over the PP analysis even in NI trials, and stated the following:

> Even in NI trials, ITT analyses have an advantage over [PP] analyses that are based on excluding from analysis the outcome information generated in patients with irregularities, since [PP] analyses fail to maintain the integrity of randomization.

He recognized that nonadherence, withdrawals from therapy, missing data, violations in entry criteria, or other deviations in the protocol often tend to reduce sensitivity to differences between regimens in an ITT analysis. Such protocol deviations are of concern, especially in NI trials, and therefore, it is important to design and conduct trials in a manner that reduces the occurrence of protocol deviations (Fleming 2008). The need to minimize protocol deviations is also expressed by Pocock (2003).

More recent articles promote the role of ITT in NI trials to the primary analysis while demoting the role of PP to supportive or secondary analysis. For example, Fleming et al. (2011, p. 436) stated the following:

> Only "as randomized" analyses, where all randomized patients are followed to their outcome or to the end of study (i.e., to the planned duration of maximum follow-up or the analysis cutoff date), preserve the integrity of the randomization and, due to their unconditional nature, address the questions of most important scientific relevance. Therefore, the preferred approach to enhancing the integrity and interpretability of the [NI] trial should be to establish performance standards for measures of quality of trial conduct (e.g., targets for enrollment and eligibility rates, event rate, adherence and retention rates, cross-in rates, and currentness of data capture) when designing the trial, and then to provide careful oversight during the trial to ensure these standards are met, with the "as randomized" analysis being primary and with analyses such as [PP] or analyses based on more sophisticated statistical models accounting for irregularities being supportive.

Another viewpoint in support of the ITT analysis is given by Schumi and Wittes (2011):

> Appeal to the dangers of sloppiness is not a reason for using the PP population, but rather is a reason for ensuring that a trial is well designed and carefully monitored, with the primary analysis performed on an ITT population.

10.6 Implementation of an Intention-to-Treat Analysis in Randomized Control Trials

10.6.1 Introduction

Since a pure ITT analysis is practically impossible, a modified ITT analysis, although not consistently defined in the literature, is often used in the analysis of RCTs. In general, there are three aspects in the analysis of an RCT:

1. The participants are grouped according to the treatment to which they were randomized (as randomized) or to the treatment that they actually received (as treated)
2. Whether or not to exclude protocol violators
3. How to deal with missing data

The ITT principle defined in Section 10.2.1 should be followed in the analysis of RCTs. Any deviation from this principle should be justified in terms of bias in the treatment comparisons and other considerations. There is no one-size-fit-all solution for a given deviation from the ITT principle. For example, subjects who were, in retrospect, ineligible for the study may or may not be excluded from the analysis, depending upon the situation (see Section 10.6.2). Sections 10.6.3 and 10.6.4 present other examples where subjects may or may not be excluded from the analysis. Sections 10.6.5 and 10.6.6 discuss missing data and noncompliance with study medication, respectively.

10.6.2 Ineligible Subjects (Inclusion/Exclusion Criteria)

In general, a study population enrolled in an RCT is defined by the inclusion/exclusion criteria. An explanatory approach recruits as homogeneous a population as possible (see Section 10.2.3). This will reduce the "noise" and thus increase the likelihood of a successful trial. Subjects enrolled in an RCT are a "convenience" sample rather than a random sample from the study population. However, to make a statistical inference regarding the study population one has to "pretend" that the enrolled subjects are a random sample. On the other hand, the study population should not be too restricted so that the trial results could be extrapolated to a more general patient population. It should be noted that such an extrapolation has to rely on subjective clinical judgments rather than on statistical inferences.

The principle for excluding subjects from the analysis without biasing the results is given by Gillings and Koch (1991) in the following:

> In general, there will be no bias for treatment comparisons if subjects are excluded based on information that could have been known prior to randomization and such exclusion is not, in any way, dependent on the treatment. It is, of course, important that there was no relationship between treatment to which the subject was randomized and the manner in which the eligibility violation was detected. For example, some patients on a new treatment under test may be reviewed more rigorously because of the scrutiny given to a new therapy. If such enhanced scrutiny led to the identification of clerical errors or other facts that in turn led to the patient violating protocol criteria, then the withdrawal of such a patient would be treatment related because only patients on test treatment received the enhanced scrutiny.

Subjects who do not meet the inclusion/exclusion criteria could be inappropriately randomized due to human error. In principle, excluding such ineligible subjects from the analysis will not lead to a biased estimate of a treatment difference. In fact, Peto et al. (1976) allow for some inappropriately randomized patients to be excluded. CPMP (1995, p. 1673) states that "Any objective entry criteria measured before randomization that will be used to exclude patients from analysis should be prespecified and justified—for example, those relating to the presence or absence of the disease under investigation." An independent adjudication committee blinded to treatment and outcome must systematically review each subject, and the decision whether to remove an ineligible subject should be based solely on information that reflects the subject's status before randomization (Fergusson et al. 2002). The ICH E9 guidance (1998, pp. 28–29) lists four requirements for excluding subjects from the ITT analysis:

1. The entry criterion was measured prior to randomization.
2. The detection of the relevant eligibility violations can be made completely objectively.
3. All subjects receive equal scrutiny for eligibility violations. (This may be difficult to ensure in an open-label study, or even in a double-blind study, if the data is unblinded prior to this scrutiny, emphasizing the importance of the blind review.)
4. All detected violations of the particular entry criterion are excluded.

The protocol should state "the intention to remove entry criteria violators and to identify the relevant criteria so that no accusations of selective analysis can later arise" (Lewis and Machin 1993). Wiens and Zhao (2007) added a fifth requirement:

5. That the criteria for excluding a subject from the ITT analysis are developed a priori to prevent post hoc subset analyses from being presented as confirmatory.

To implement this requirement, they suggested "to form an independent subject eligibility committee composed of people not involved in the conduct of the study" (Wiens and Zhao 2007).

There are occasions when the exclusion of entry criteria violators is important. Such a scenario is discussed by Lewis and Machin (1993, p. 649) in the following:

> Suppose that drug X has been shown to be effective in mild and moderate disease, but there remains a question as to whether it is effective in severe cases, and so a trial in severe patients is started. Suppose also that some patients with moderate disease are inadvertently entered into this trial. Failure to exclude these patients from the analysis would bias the results in [favor] of drug X, or at any rate would lead to the answering of the wrong question.

On the other hand, there are exceptions to the exclusions of entry criteria violators. One such exception is given by Gillings and Koch (1991, pp. 415–416) as follows:

> For example, subjects with a prior history of cardiovascular problems or women not using effective birth control are excluded from a particular trial for safety reasons. … In general, patients who should have been excluded for safety reasons but who do participate in the trial may not be appropriate candidates for exclusion from analysis because the original safety reason for exclusion, once violated, can be irrelevant from an efficacy perspective.

Newell (1992) recommends that randomization be done as late as possible when diagnostic and consent procedures have all been completed (see Figure 10.2 in Section 10.3.1). This helps reduce entry criteria violators to a reasonable minimum. However, it is not always possible to do so (see Example 4 in Section 10.6.4).

10.6.3 Subjects Who Do Not Receive the Randomized Treatment

In an RCT, subjects may or may not receive the randomized treatment, or may be switched to the other treatment for various reasons. As stated by Gillings and Koch (1991, p. 413), "... analyzing subjects as though they were given a treatment that they did not actually receive is typically regarded as inappropriate from the perspective of clinical medicine." An alternative approach is (1) to exclude such subjects, as in the PP analysis (see Section 10.3.3), or (2) to include such subjects according to the treatment actually received, as in the AT analysis (see Section 10.3.4). Depending upon the reasons for not receiving the randomized treatment for a given study design, the PP or AT analysis may or may not introduce bias in the treatment comparison. Some examples are discussed in the following paragraphs.

Example 1a: Randomized treatment is not needed due to premature randomization

This example is discussed by Fergusson et al. (2002, p. 654) as follows:

> In one trial of leucodepletion of red blood cells, patients were randomized before an operation rather than when a unit of red blood cells was requested by the surgical team (Houbiers et al. 1994). The point of randomization was premature, and 36% of the patients randomized to the study did not need a transfusion. ... Excluding all randomized patients who did not receive a unit of red blood cells will not bias the analysis, as long as allocation to treatment or control arm could not influence the likelihood that patients receive a transfusion.

Excluding all such patients actually enhances the precision of the estimate and provides a meaningful estimate of relative risk reduction for the clinician (Fergusson et al. 2002). Fergusson et al. (2002, p. 654) recommend that "investigators should report an analysis of all randomized patients, as well as baseline characteristics for all patients excluded from the analysis."

Example 1b: Randomized treatment is not needed in the treatment arm only

This example is discussed by Fergusson et al. (2002, p. 654) as follows:

> In studies in which only patients allocated to one of two arms will receive the target intervention, excluding such patients will lead to biased results. For example, in a clinical trial of epidural anesthesia in childbirth, some women randomized to the epidural treatment arm did not need an epidural because their pain levels did not rise above their personal thresholds

[Loughnan et al. 2000]. Investigators should not exclude these patients from the analysis, as they cannot identify similar patients in the control arm.

Example 2: Hypothetical example comparing medical versus surgical treatment as described by Newell (1992)

"Suppose, for dramatic simplicity, that patients are randomly allocated to medical or surgical treatment. Those allocated to medical treatment are given medication immediately, while those allocated to surgery may require preparation, possibly waiting a few days or weeks for an available surgical [theater] time-slot. If a patient should happen to die before reaching the operating theater, a surgeon might be inclined to say 'that death should not count against the surgical option—I didn't get a chance to put my knife into the patient.' The physician would rightly claim that if the surgeon could discount these—obviously the sickest—patients, the comparison would not be fair. In fact, the surgical [program] includes some (inevitable) delays, and all mortality occurring after the decision to perform surgery must correctly be assigned as part of the outcome of that [program]."

Example 3: Subjects do not receive the randomized treatment due to administrative errors

Gillings and Koch (1991, p. 413) recommended that subjects be analyzed according to the treatments actually received "under the circumstance that only a few patients were wrongly randomized and that such administrative errors were not associated with the background characteristics of these patients or their prognosis." They noted that in a large multicenter trial, there is likely at least one randomization error no matter how carefully the study procedures are implemented.

10.6.4 Subjects Who Do Not Have the Condition of Interest or the Underlying Disease

Subjects who were randomized might not have the condition of interest or the underlying disease either (1) because they were randomized before eligibility for inclusion could be confirmed (Example 4) or (2) due to poor or excessively broad eligibility criteria (Example 5). Subjects who do not have the condition of interest or the underlying disease should not be excluded from the ITT analysis using the pragmatic approach for effectiveness; however, such subjects should be excluded from the analysis using the explanatory approach for efficacy (see Sections 10.2.2 and 10.2.3). Such an analysis may be called a PP analysis in Example 4. However, it is not clear what analysis it should be called in Example 5 because such subjects actually meet the eligibility criteria.

Example 4: Subjects randomized before eligibility for inclusion can be confirmed

If investigators expect delays in obtaining clinical or laboratory information on subjects' eligibility, they should ideally postpone randomization until this information is available. However, even with sound methods and procedures and the best of intentions instances will occur when subjects must be randomized before all the data needed to confirm eligibility is available.

In a study of an anti-influenza drug, oseltamivir carboxylate (Treanor et al. 2000), all consenting subjects who present to a doctor within 48 hours of development of influenzalike symptoms are enrolled and randomized into the trial. The study protocol stipulates that only subjects who later give positive results on culture or serological tests for influenza infection will be included in the analysis. Of the 629 subjects who were randomized, 255 (40%) were later found to not have influenza. The study reported that in the 374 patients who were infected, the study drug reduced the duration of illness by 30% (p <0.001). However, analysis of all 629 randomized patients shows a less dramatic but still significant effect of the study drug, with a reduced duration of 22% (p = 0.004).

Fergusson et al. (2002, p. 653) states the following:

> This clinical scenario mirrors real-life clinical situations where doctors need to treat patients before all information is available. The major issue in the interpretation of results becomes one of effectiveness versus efficacy [see Section 10.2.2] or explanatory approach versus pragmatic approach [see Section 10.2.3]. One would want to be sure that the benefit of the study drug to patients with the underlying condition outweighs the harm to patients exposed to the drug without possibility of benefit. Therefore, the primary presentation of the results should include all patients randomized into the study.

Example 5: Poor or excessively broad eligibility criteria

This example is discussed by Fergusson et al. (2002, p. 653) as follows:

> Poorly defined or excessively broad eligibility criteria can lead to the inclusion of patients who do not have the condition of interest and are, therefore, unlikely to benefit from treatment. For example, studies of severe infections resulting in sepsis syndrome are often beset by difficulties in defining the condition of interest and the eligibility criteria [Bone et al. 1992; Cohen et al. 2001]. The diversity of clinical presentations often results in the enrollment of patients who meet eligibility criteria and receive treatment but are unlikely to benefit. ... A large [RCT] of a drug that modulates immune responses in severe sepsis enrolled a very diverse study population because of broad eligibility criteria [Fisher et al. 1994]. A high proportion (175/893 or 20%) of enrolled patients that met the criteria did not have a confirmed infection, resulting in a study that yielded a less-than-optimal test of the researchers' hypothesis. ... Under such circumstances, the primary analysis should include all randomized patients. A secondary analysis that includes only patients who had the condition of interest and that is based on data collected before randomization can also be informative and unbiased.

10.6.5 Missing Data

Since missing data is unavoidable in clinical trials with human subjects, to include all randomized subjects in the ITT analysis, one has to resort to data imputation. However, such methods require an untestable assumption of missing at random to some degree. The best way to deal with the problem is to have as little missing data as possible (Lachin 2000; NRC 2010) (see Section 10.3.2). In this regard, there is a need for a cultural shift, focusing on strategies to prevent missing data during the conduct and management of clinical trials, rather than relying on imperfect analytic methods (O'Neill and Temple 2012; Dziura et al. 2013). How to deal with missing data in the analysis is beyond the scope of this book. The reader is referred to Rothmann, Wiens, and Chan (2012).

At the request of the U.S. Food and Drug Administration (FDA), the Panel on the Handling of Missing Data in Clinical Trials was created by the National Research Council's Committee on National Statistics. The panel's work focused primarily on phase 3 confirmatory clinical trials that are the basis for the approval of drugs and devices. The panel concluded that a more principled approach to design and analysis in the presence of missing data is both needed and possible. Such an approach needs to focus on two critical elements: (1) careful design and conduct to limit the amount and impact of missing data, and (2) analysis that makes full use of information on all randomized participants and is based on careful attention to assumptions about the nature of the missing data underlying estimates of treatment effects. The panel presented 18 recommendations in the prevention and treatment of missing data in clinical trials (NRC 2010) (see Appendix 10.A).

Li et al. (2014) performed a systematic review on the prevention and handling of missing data for patient-centered outcomes research (PCOR). PCOR focuses on comparative effectiveness research that "helps people and their caregivers communicate and make informed health-care decisions, allowing their voices to be heard in assessing the value of health-care options" (PCOR, 2014). Note that the PCOR could be RCTs or observational studies, such as cohort, case-control, and cross-sectional studies.

The authors identified 30 guidance documents that include at least one formal recommendation about how missing data should be prevented or handled. These 30 documents were published between 1996 and 2011 (5 were in draft form), with more than half published after 2008, and 12 of which were written for RCTs. Almost one-third (9 of 30, 30%) were prepared by the International Society for Pharmacoeconomics and Outcomes Research (ISPOR), followed by 5 of 30 (17%) prepared by the FDA and 4 of 30 (13%) by the Expert Working Group (Efficacy) of the International Conference on Harmonization of Technical Requirements for Registration of Pharmaceuticals for Human Use (ICH).

The authors extracted 39 recommendations on the prevention and handling of missing data. The two-round consensus process and discussion yielded 10 mandatory standards: 3 on study design, 2 on conduct, 3 on analysis, and 2 on reporting (see Appendix 10.B). The four domains (i.e., study design, conduct, analysis, and reporting) are in line with the recommendation by Hollis and Campbell (1999) (see Section 10.4.2).

10.6.6 Noncompliance with Study Medication

Showing benefit in an explanatory trial does not necessarily lead to clinical benefits in practice using design (1) discussed in Section 10.2.3. In principle, from regulatory and practical points of view, medical intervention should be shown to provide clinical benefits in a real-world setting—that is, showing effectiveness rather than efficacy. See Section 10.2.2 for the definitions of effectiveness and efficacy. Therefore, noncompliant subjects should be included in the ITT analysis, and no adjustment for noncompliance is necessary. However, it should be noted that excessive noncompliance could render the results uninterpretable. See Section 10.4.3 for further discussion on noncompliance.

10.7 Discussion

How much missing data is too much? Schulz and Grimes (2001) suggest that losses to follow-up less than 5% usually have little impact, whereas losses greater than 20% raise serious issues about study validity. In-between levels lead to problems somewhere in the middle. To support this, Kristman, Manno, and Cote (2004) demonstrated through simulation that substantial bias in the estimation of odds ratios under missing not at random (MNAR) conditions may arise in cohort studies, with loss to follow-up of 20%. However, the 5-20 rule of thumb has no statistical justification and oversimplifies the problem, as the bias resulting from missing data also depends on the missing data mechanism and the analytic method (Dziura et al. 2013).

Gillings and Koch (1991) suggest a threshold value of 5% in the context of practically defining the ITT population: "All patients randomized who were known to take at least one dose of treatment and who provided any follow-up data for one or more key efficacy variables; in turn, ITT patients are allocated to treatments actually received." They suggest that randomization errors and attrition through not taking at least one dose and not providing any follow-up data should be kept to 5%. It should be noted, however, that subjects who did not receive the surgical procedure in Example 2 in Section 10.6.3 should not be excluded. Therefore, it is important to realize that each therapeutic area may have its own unique issues, as noted in Section 10.1, so that recommendations for a given therapeutic area may not be applicable to another therapeutic area.

Deviations from the ITT principle (i.e., subjects not being analyzed as randomized or post-randomization exclusion of subjects) may or may not be acceptable, depending upon the situation, as discussed in Sections 10.6.2, 10.6.3, and 10.6.4, although they are limited in scope. Tremendous effort is needed to research examples in the literature, covering more scenarios for a wider range of therapeutic areas.

Hopefully, recent research efforts in the prevention and treatment of missing data in clinical trials (e.g., NRC 2010; Dziura et al. 2013; Li et al. 2014) will reduce the amount of missing data and improve study quality in the near future. With improved quality of the trial conduct, it is hopeful that the number of protocol violators is minimal, as is the risk of falsely declaring NI using ITT analysis in settings where T truly is clinically inferior to S.

Appendix 10.A Eighteen Recommendations by the National Research Council

Trial objectives

1. The trial protocol should explicitly define (1) the objective(s) of the trial; (2) the associated primary outcome or outcomes; (3) how, when, and on whom the outcome or outcomes will be measured; and (4) the measures of intervention effects, that is, the causal estimands of primary interest. These measures should be meaningful for all study participants and estimable with minimal assumptions. Concerning the latter, the protocol should address the potential impact and treatment of missing data.

Reducing dropouts through trial design

2. Investigators, sponsors, and regulators should design clinical trials consistent with the goal of maximizing the number of participants who are maintained on the protocol-specified intervention until the outcome data is collected.

3. Trial sponsors should continue to collect information on key outcomes on participants who discontinue their protocol-specified intervention in the course of the study, except in those cases for which a compelling cost-benefit analysis argues otherwise, and this information should be recorded and used in the analysis.

4. The trial design team should consider whether participants who discontinue the protocol intervention should have access to and be encouraged to use specific alternative treatments. Such treatments should be specified in the study protocol.

5. Data collection and information about all relevant treatments and key covariates should be recorded for all initial study participants, whether or not participants received the intervention specified in the protocol.

Reducing dropouts through trial conduct

6. Study sponsors should explicitly anticipate potential problems of missing data. In particular, the trial protocol should contain a section that addresses missing data issues, including the anticipated amount of missing data and steps taken in trial design and trial conduct to monitor and limit the impact of missing data.

7. Informed consent documents should emphasize the importance of collecting outcome data from individuals who choose to discontinue treatment during the study, and they should encourage participants to provide this information whether or not they complete the anticipated course of study treatment.

8. All trial protocols should recognize the importance of minimizing the amount of missing data, and, in particular, they should set a minimum rate of completeness for the primary outcome(s) based on what has been achievable in similar past trials.

Treating missing data

9. Statistical methods for handling missing data should be specified by clinical trial sponsors in study protocols and their associated assumptions stated in a way that can be understood by clinicians.

10. Single-imputation methods, such as last observation carried forward and baseline observation carried forward, should not be used as the primary approach to the treatment of missing data unless the assumptions that underlie them are scientifically justified.

11. Parametric models in general, and random-effects models in particular, should be used with caution, with all their assumptions clearly spelled out and justified. Models relying on parametric assumptions should be accompanied by goodness-of-fit procedures.

12. It is important that the primary analysis of the data from a clinical trial should account for the uncertainty attributable to missing data so that under the stated missing data assumptions, the associated significance tests have valid Type I error rates and the confidence intervals have the nominal coverage properties. For inverse probability weighting and maximum likelihood methods, this analysis can be accomplished by appropriately computing standard errors, using either asymptotic results or the bootstrap method. It is necessary to use appropriate rules for multiplying imputing missing responses and combining results across imputed datasets, because single imputation does not account for all sources of variability.

13. Weighted generalized estimating equation methods should be more widely used in settings where missing at random can be well justified and a stable weight model can be determined as a possibly useful alternative to parametric modeling.

14. When substantial missing data is anticipated, auxiliary information should be collected that is believed to be associated with reasons for missing values and with the outcomes of interest. This could improve the primary analysis through use of a more appropriate missing-at-random model or help carry out sensitivity analyses to assess the impact of missing data on estimates of treatment differences. In addition, investigators should seriously consider following up with all or a random sample of trial dropouts who have not withdrawn consent to ask them why they dropped out of the study and, if they are willing, to collect outcome measurements from them.

15. Sensitivity analyses should be part of the primary reporting of findings from clinical trials. Examining sensitivity to the assumptions about the missing data mechanism should be a mandatory component of reporting.

Understanding the causes and degree of dropouts in clinical trials

16. The FDA and the National Institutes of Health should make use of their extensive clinical trial databases to carry out a program of research, both internal and external, to identify common rates and causes of missing data in different domains and how different models perform in different settings. The results of such research can be used to inform future study designs and protocols.

17. The FDA and drug, device, and biologic companies that sponsor clinical trials should carry out continued training of their analysts to keep abreast on up-to-date techniques for missing data analysis. The FDA should also encourage continued training of their clinical reviewers to make them broadly familiar with missing data terminology and missing data methods.

18. The treatment of missing data in clinical trials, being a crucial issue, should have a higher priority for sponsors of statistical research, such as the National Institutes of Health and the National Science Foundation. Progress is particularly needed in several important areas, namely (1) methods for sensitivity analysis and principled decision making based on the results from sensitivity analyses, (2) analysis of data where the missingness pattern is nonmonotone, (3) sample size calculations in the presence of missing data, and (4) design of clinical trials—in particular, plans for follow-up after treatment discontinuation (degree of sampling, how many attempts were made, etc.)—and (5) doable robust methods to more clearly understand their strengths and vulnerabilities in practical settings. The development of software that supports coherent missing data analyses is also a high priority.

Appendix 10.B Minimum Standards in the Prevention and Handling of Missing Data in Patient-Centered Outcomes Research

Standards on study design

1. Define [the] research question, in particular, the outcome(s):

- The study protocol should explicitly define (1) the objective(s) of the study; (2) the intervention(s) of interest; (3) the associated primary outcome(s) that quantify the impact of interventions for a defined period; (4) how, when, and on whom the outcome(s) will be measured; (5) potential confounders, if relevant; and (6) the measures of intervention effects—that is, the parameters (estimands) that capture the causal effect of the intervention of primary interest. The parameters should be meaningful for all study participants and estimable with minimal assumptions. This standard applies to all study designs that aim to assess intervention effectiveness.

- Defining outcome(s) precisely and accurately requires careful attention, because the choice of outcome may have important implications for study design, implementation, expected amount of and reason for missing data, and methods for handling missing data. For example, the outcome could be defined in the population that includes all participants randomized to the study intervention(s), regardless of the intervention participants actually received (i.e., ITT estimand), or the outcome could be defined in a more restricted population that includes only those who can tolerate the intervention for a given period. Outcome(s) could be measured after a short or long follow-up period and measured at one point in time or repeatedly over time. At a minimum, the primary outcome must be decided upon and adequately described in the study protocol. Imprecise and vague definition may lead to a lack of clarity in how to prevent and handle missing data.

2. Take steps in design and conduct to minimize missing data:

- Investigators should explicitly anticipate potential problems of missing data. The study protocol should contain a section that addresses missing data issues and steps taken in study design and conduct to monitor and limit the impact of missing data. If relevant, the protocol should include the anticipated amount of and reasons for missing data and plans to follow up with participants. This standard applies to all study designs for any type of research question.

3. Prespecify statistical methods for handling missing data:

- Statistical methods for handling missing data should be prespecified in the study protocol, and their associated assumptions stated in a way that can be understood by all stakeholders. The reasons for missing data should be considered in the analysis. This standard applies to all study designs for any type of research question.

Standards on study conduct

4. Continue collecting information on key outcomes:

- Whenever a participant discontinues some or all types of participation in a research study, the investigator should document the following: (1) the reason for discontinuation, (2) who decided that the participant would discontinue, and (3) whether the discontinuation involves some or all types of participation. Investigators should continue to collect information on key outcomes for participants who discontinue the protocol-specified intervention. This standard applies to prospective study designs that aim to assess intervention effectiveness.

5. Monitor missing data:

- For studies that include a data and safety monitoring board, the board should review plans for and the implementation of the prevention and handling of missing data. The board should review completeness and timeliness of data and recommend modifications as appropriate.

Standards on analysis

6. Account for uncertainty in handling missing data in the analysis:

- Statistical inference of intervention effects or measures of association should account for statistical uncertainty attributable to missing data. This means that under the stated missing data assumptions of the methods used for imputing missing data, the associated significance tests should have valid Type I error rates and confidence intervals should have the nominal coverage properties. This standard applies to all study designs for any type of research question.

7. Discourage single-imputation methods:

- Single-imputation methods, such as the last observation carried forward and baseline observation carried forward, generally should not be used as the primary approach for handling missing data in the analysis. This standard applies to all study designs for any type of research question.

8. Conduct sensitivity analysis:

- Examining sensitivity to the assumptions about the missing data mechanism (i.e., sensitivity analysis) should be a mandatory component of the study protocol, analysis, and reporting. This standard applies to all study designs for any type of research question.

Standards on reporting

9. Account for all participants entered in the study when reporting the results:

- All participants who enter the study should be accounted for when reporting the results, whether or not they are included in the analysis. Describe and justify any planned reasons for excluding participants from analysis. This standard applies to all study designs for any type of research question.

10. Report on data completeness and strategies applied to handle missing data:

- Report on data completeness and how missing data was handled in the analysis to facilitate interpretation of study results. The potential influence of missing data on the study results should be described. This standard applies to all study designs for any type of research question.

References

Alshurafa M, Briel M, Akl EA, Haines T, Moayyedi P, Gentles SJ, Rios L, Tran C, Bhatnagar N, Lamontagne F, Walter SD, and Guyatt GH (2012). Inconsistent Definitions for Intention-to-Treat in Relation to Missing Outcome Data: Systematic Review of the Methods Literature. PLoS ONE 7(11): e49163. doi:10.1371/journal.pone.0049163. http://www.plosone.org/article/info%3Adoi%2F10.1371%2Fjournal.pone.0049163 (Accessed: September 9, 2013).

Bone RC, Balk RA, Cerra FB, Dellinger RP, Fein AM, Knaus WA, Schein RMH, and Sibbald WJ (1992). Definitions for Sepsis and Organ Failure and Guidelines for the Use of Innovative Therapies in Sepsis. The ACCP/SCCM Consensus Conference Committee. American College of Chest Physicians/Society of Critical Care Medicine. *Chest*, **101**:1644–1655.

Bradford-Hill A (1961). *Principles of Medical Statistics*. New York, NY: Oxford University Press.

Brittain E and Lin D (2005). A Comparison of Intent-to-Treat and Per-Protocol Results in Antibiotic Noninferiority Trials. *Statistics in Medicine*, **24**:1–10.

Cochrane Collaboration (2002). Module 14: Further Issues in Meta-Analysis (Cochrane Handbook for Systematic Reviews of Interventions). http://www.cochrane-net.org/openlearning/PDF/Module_14.pdf (Accessed: September 10, 2013).

Cohen J, Guyatt G, Bernard GR, Calandra T, Cook D, Elbourne D, Marshall J, Nunn A, and Opal S (2001), for a UK Medical Research Council International Working Party. New Strategies for Clinical Trials in Patients with Sepsis and Septic Shock. *Critical Care Medicine Journal*, **29**:880–886.

CPMP Note for Guidance (III/3630/92-EN; 1995). Biostatistical Methodology in Clinical Trials in Applications for Marketing Authorizations for Medicinal Products. *Statistics in Medicine*, **14**:1659–682.

D'Agostino RB Sr, Massaro JM, and Sullivan LM (2003). Non-inferiority Trials: Design Concepts and Issues: The Encounters of Academic Consultants in Statistics. *Statistics in Medicine*, **22**:169–186.

Dziura JD, Post LA, Zhao Q, Fu Z, and Peduzzi P (2013). Strategies for Dealing with Missing Data in Clinical Trials: From Design to Analysis. *Yale Journal of Biology and Medicine*, **86**:343–358.

Ebbutt AF and Frith L (1998). Practical Issues in Equivalence Trials. *Statistics in Medicine*, **17**:1691–1701.

European Agency for the Evaluation of Medicinal Products, Committee for Proprietary Medicinal Products (2000). Points to Consider on Switching Between Superiority and Noninferiority. http://www.ema.europa.eu/docs/en_GB/document_library/Scientific_guideline/2009/09/WC500003658.pdf (Accessed: August 25, 2013).

Fergusson D, Aaron S, Guyatt G, and Herbert P (2002). Post-randomisation Exclusions: The Intention-to-Treat Principle and Excluding Patients from Analysis. *British Medical Journal*, **325**:652–654.

Fisher CJ Jr, Dhainaut JF, Opal SM, Pribble JP, Balk RA, Slotman GJ, Iberti TJ, Rackow EC, Shapiro MJ, Greenman RL, Reines HD, Shelly MP, Thompson BW, LaBrecque JF, Michael A, Catalano MA, Knaus WA, and Sadoff JC (1994). Recombinant Human Interleukin 1 Receptor Antagonist in the Treatment of Patients with Sepsis Syndrome. Results from a Randomized, Double-Blind, Placebo-Controlled Trial. Phase III rhIL1-ra Sepsis Syndrome Study Group. *Journal of American Medical Association*, **271**:1836–1843.

Fisher LD, Dixon DO, Herson J, Frankowski RK, Hearron MS, and Peace KE (1990). Intention to Treat in Clinical Trials, in *Statistical Issues in Drug Research and Development* (American Statistical Associations Group), ed Peace K E, Marcel Dekker. New York: 331–350.

Fleming TR (2008). Current Issues in Non-inferiority Trials. *Statistics in Medicine*, **27**:317–332.

Fleming TR, Odem-Davis K, Rothmann MD, and Shen YL (2011). Some Essential Considerations in the Design and Conduct of Non-inferiority Trials. *Clinical Trials*, **8**:432–439.

Garrett AD (2003). Therapeutic Equivalence: Fallacies and Falsification. *Statistics in Medicine*, **22**:741–762.

Gillings D and Koch G (1991). The Application of the Principle of Intention-to-Treat to the Analysis of Clinical Trials. *Drug Information Journal*, **25**:411–424.

Gonzalez CD, Bolaños R, and de Sereday M (2009). Editorial on Hypothesis and Objectives in Clinical Trials: Superiority, Equivalence and Non-inferiority, *Thrombosis Journal*, **7**:3 doi: 10.1186/1477-9560-7-3.

Gøtzsche PC (2006). Lessons from and Cautions about Noninferiority and Equivalence Randomized Trials [Editorial]. *Journal of the American Medical Association*, **295**:1172–1174.

Gravel J, Opartny L, and Shapiro S (2007). The Intention-to-Treat Approach in Randomized Trials: Are Authors Saying What They Do and Doing What They Say? *Clinical Trials*, **4**:350–356.

Greene CJ, Morland LA, Durkalski VL, and Frueh BC (2008). Noninferiority and Equivalence Designs: Issues and Implications for Mental Health Research, *Journal of Traumatic Stress*, **21**(5):433–443.

Hauck WW, Anderson S. (1999). Some Issues in the Design and Analysis of Equivalence Trials. *Drug Information Journal*, **33**:109–118.

Heritier SR, Gebski VJ, and Keech AC (2003). Inclusion of Patients in Clinical Trial Analysis: The Intention-to-Treat Principle. *Medical Journal of Australia*, **179**:438–440.

Hernan MA and Hernandez-Diaz S (2012). Beyond the Intention-to-Treat in Comparative Effectiveness Research. *Clinical Trials*, **9**(1):48–55.

Hill CL, LaValley MP, and Felson DT (2002). Secular Changes in the Quality of Published Randomized Clinical Trials in Rheumatology. *Arthritis Rheum*, **46**:779–784.

Hollis S and Campbell F (1999). What Is Meant by Intention-to-Treat Analysis? Survey of Published Randomised Controlled Trials. *British Medical Journal*, **319**(7211):670–674.

Houbiers JG, Brand A, van de Watering LM, Hermans J, Verwey PJ, Bijnen AB, Pahlplatz P, Schattenkerk ME, Wobbes T, de Vries JE, Klementschitsch P, van de Maas AHM, and van de Velde CJH (1994). Randomised Controlled Trial Comparing Transfusion of LeucocyteDepleted or Buffy Coat Depleted Blood in Surgery for Colorectal Cancer. *Lancet*, **344**:573–578.

International Conference on Harmonization (ICH) E9 Guideline (1998). *Statistical Principles for Clinical Trials*. http://www.fda.gov/downloads/Drugs/GuidanceComplianceRegulatoryInformation/Guidances/UCM073137.pdf (Accessed: September 27, 2012).

Kaul S and Diamond GA (2007). Making Sense of Noninferiority: A Clinical and Statistical Perspective on Its Application to Cardiovascular Clinical Trials. *Progress in Cardiovascular Diseases*, **49**(4):284–299.

Kristman V, Manno M, and Cote P (2004). Loss to Follow-up in Cohort Studies: How Much Is Too Much? *European Journal of Epidemiology*, **19**(8):751–760.

Lachin JM (2000). Statistical Considerations in the Intent-to-Treat Principle. *Controlled Clinical Trials*, **21**(3):167–189.

Le Henanff A, Giraudeau B, Baron G, and Ravaud P (2006). Quality of Reporting of Noninferiority and Equivalence Randomized Trials. *Journal of the American Medical Association*, **295**:1147–1151.

Lesaffre E (2008). Superiority, Equivalence and Non-inferiority Trials. *Bulletin of the NYU Hospital for Joint Diseases*, **66**(2):150–154.

Lewis JA and Machin D (1993). Intention to Treat: Who Should Use ITT? *British Journal of Cancer*, **68**:647–650.

Li T, Hutfless S, Scharfstein DO, Daniels MJ, Hogan JW, Little RJA, Roy JA, Law AH, and Dickersin K (2014). Standards Should Be Applied in the Prevention and Handling of Missing Data for Patient-Centered Outcomes Research: A Systematic Review and Expert Consensus. *Journal of Clinical Epidemiology*, **67**:15–32.

Loughnan BA, Carli F, Romney M, Dore CJ, and Gordon H (2000). Randomized Controlled Comparison of Epidural Bupivacaine versus Pethidine for Analgesia in Labour. *British Journal of Anaesthesia*, **84**:715–719.

Moher D, Hopewell S, Schulz KF, Montori V, Gotzsche PC, Devereaux PJ, Elbourne D, Egger M, and Altman DG (2010). CONSORT 2010 Explanation and Elaboration: Updated Guidelines for Reporting Parallel Group Randomised Trials. *British Medical Journal*, **340**:c869.

Moher D, Schulz KF, and Altman DG (2001). The CONSORT Statement: Revised Recommendations for Improving the Quality of Reports of Parallel-Group Randomized Trials. *Annals of Internal Medicine*, **134**:657–662.

National Research Council (2010). The Prevention and Treatment of Missing Data in Clinical Trials. Panel on Handling Missing Data in Clinical Trials. Committee on National Statistics, Division of Behavioral and Social Sciences and Education. Washington, DC: The National Academies Press. http://csph.ucdenver .edu/sites/kittelson/Bios6648-2013/Lctnotes/2013/NASmissingData.pdf (Accessed: January 5, 2014).

Neal K (2009). Efficacy vs. Effectiveness. http://getedited.wordpress.com /2009/10/26/efficacy-vs-effectiveness (Accessed: November 3, 2013).

Newell DJ (1992). Intention-to-Treat Analysis: Implications for Quantitative and Qualitative Research. *International Journal of Epidemiology*, **21**:837–841.

Ng T-H (2001). Choice of delta in equivalence testing, *Drug Information Journal*, **35**:1517–1527.

Nordic Council on Medicine (1989). *Good Clinical Practice*. Uppsala, Sweden: Nordic Guidelines, NLN Publication No. 28.

O'Neill RT and Temple R (2012). The Prevention and Treatment of Missing Data in Clinical Trials: An FDA Perspective on the Importance of Dealing with It. *Journal of Clinical Pharmacy and Therapeutics*, **91**(3):550–554.

Pater C (2004). Equivalence and Noninferiority Trials: Are They Viable Alternatives for Registration of New Drugs? (III). *Current Controlled Trials in Cardiovascular Medicine*, **5**:8.

Peto R, Pike MC, Armitage P, Breslow NE, Cox DR, Howard SV, Mantel N, McPherson K, Peto J, and Smith PG (1976). Design and Analysis of Randomized Clinical Trials Requiring Prolonged Observation of Each Patient. I. Introduction and Design. *British Journal of Cancer*, **34**:585–612.

Piaggio G, Elbourne DR, Altman DG, Pocock SJ, and Evans SJ (2006). Reporting of Noninferiority and Equivalence Randomized Trials: An Extension of the CONSORT Statement. *Journal of American Medical Association*, **295**:1152–1160.

Pocock SJ (2003). Pros and Cons of Noninferiority Trials. *Blackwell Publishing Fundamental & Clinical Pharmacology*, **17**:483–490.

Polit DF, and Gillespie B (2009). The Use of the Intention-to-Treat Principle in Nursing Clinical Trials. *Nursing Research*, **59**:391–399.

Polit DF, and Gillespie BM (2010). Intention-to-Treat in Randomized Controlled Trials: Recommendations for a Total Trial Strategy. *Research in Nursing & Health*, **33**(4):355–368.

Roland M, and Torgerson DJ (1998). Understanding Controlled Trials. What Are Pragmatic Trials? *British Medical Journal*, **316**:285.

Rothmann MD, Wiens BL, and Chan ISF (2011). *Design and Analysis of Non-Inferiority Trials*. Boca Raton, FL: Chapman & Hall/CRC.

Sanchez MM and Chen X (2006). Choosing the Analysis Population in Non-inferiority Studies: Per Protocol or Intent-to-Treat. *Statistics in Medicine*, **25**:1169–1181.

Schulz KF, Altman DG, and Moher D (2010). CONSORT 2010 Statement: Updated Guidelines for Reporting Parallel Group Randomised Trials. *British Medical Journal*, **340**:c332.

Schulz KF, and Grimes D (2002). Sample Size Slippages in Randomised Trials: Exclusions and the Lost and Wayward. *Lancet*, **359**(9308):781–785.

Schumi J and Wittes JT (2011). Through the Looking Glass: Understanding Non-inferiority, *Trials*, **12**:106.

Schwartz D and Lellouch J (1967). Explanatory and Pragmatic Attitudes in Therapeutic Trials. *Journal of Chronic Diseases*, **20**:637–648.

Scott IA (2009). Non-inferiority Trials: Determining Whether Alternative Treatments Are Good Enough. *Medical Journal of Australia*, **190**: 326–330.

Sheng D and Kim MY (2006). The Effects of Non-compliance on Intent-to-Treat Analysis of Equivalence Trials. *Statistics in Medicine*, **25**:1183–1199.

Thaul S (2012). How FDA Approves Drugs and Regulates Their Safety and Effectiveness. CRS (Congressional Research Service) Report R41983 2012. http://www.fas.org/sgp/crs/misc/R41983.pdf (Accessed: November 3, 2013).

Treadwell J, Uhl S, Tipton K, Singh S, Santaguida L, Sun X, Berkman N, Viswanathan M, Coleman C, Shamliyan T, Wang S, Ramakrishnan R, and Elshaug A (2012). Assessing Equivalence and Noninferiority. Methods Research Report. (Prepared by the EPC Workgroup under Contract No. 290-2007-10063.) AHRQ Publication No. 12-EHC045-EF. Rockville, MD: Agency for Healthcare Research and Quality, June 2012. http://www.effectivehealthcare.ahrq.gov/ehc/products/365/1154/Assessing-Equivalence-and-Noninferiority_FinalReport_20120613.pdf (Accessed: March 1, 2014).

Treanor JJ, Hayden FG, Vrooman PS, Barbarash R, Bettis R, Riff D, Kinnersley N, Ward P, and Mills RG. Efficacy and Safety of the Oral Neuraminidase Inhibitor Oseltamivir in Treating Acute Influenza: A Randomized Controlled Trial. *Journal of the American Medical Association*, **283**:1016–1024.

U.S. Food and Drug Administration (1988). *Guideline for the Format and Content of the Clinical and Statistical Sections of New Drug Applications*. Rockville, MD: U.S. Department of Health and Human Services.

U.S. Food and Drug Administration (2010). *Draft Guidance for Industry: Non-inferiority Clinical Trials*. http://www.fda.gov/downloads/Drugs/GuidanceComplianceRegulatoryInformation/Guidances/UCM202140.pdf (Accessed: August 25, 2013).

Wangge G, Klungel OH, Roes KCB, de Boer A, Hoes AW, et al. (2010). Room for Improvement in Conducting and Reporting Non-Inferiority Randomized Controlled Trials on Drugs: A Systematic Review. PLoS ONE 5(10): e13550. doi:10.1371/journal.pone.0013550.

Wertz RT (1995). Intention to Treat: Once Randomized, Always Analyzed. *Clinical Aphasiology*, **23**:57–64.

Wiens BL and Zhao W (2007). The Role of Intention to Treat in Analysis of Noninferiority Studies. *Clinical Trials*, **4**:286–291.

11

Thrombolytic Example

11.1 Introduction

In 1999, the Center for Biologics Evaluation and Research (CBER) of the U.S. Food and Drug Administration (FDA) determined that for phase 3 studies, "a result where the 95% one-sided confidence interval of the relative risk of mortality of the new thrombolytic [versus an accelerated tissue plasminogen activator regimen (tPA90)] that excludes a value of 1.143 provides evidence that the new thrombolytic agent has retained sufficient efficacy to be supportive of a marketing application" (FDA/CBER Memorandum 1999). In other words, the null hypothesis of interest is given by Equation 3.2 in Section 3.4 of Chapter 3 (adapted to relative risk), where r = 1.143; that is,

$$H_0: T/S \geq r = 1.143$$

versus

$$H_1: T/S < r = 1.143$$

or

$$H_0: (\text{new thrombolytic})/\text{tPA90} \geq r = 1.143$$

versus

$$H_1: (\text{new thrombolytic})/\text{tPA90} < r = 1.143$$

The main purpose of this chapter is to revisit the determination of 1.143, as opposed to presenting up-to-date information regarding the thrombolytic therapies for acute myocardial infarction.

11.2 Background

11.2.1 Thrombolytic and Fibrinolytic Therapies for Acute Myocardial Infarction

This subsection introduces the medical term of heart attack and a brief history of its treatment.

> Myocardial infarction (MI) or acute myocardial infarction (AMI) is the medical term for an event commonly known as a heart attack. It happens when blood stops flowing properly to part of the heart and the heart muscle is injured due to not receiving enough oxygen. Usually, this is because one of the coronary arteries that supplies blood to the heart develops a blockage due to an unstable buildup of white blood cells, cholesterol, and fat. The event is called "acute" if it is sudden and serious (Wikipedia Contributors 2014a).

"Thrombolytic therapy is the use of drugs to break up or dissolve blood clots, which are the main cause of both heart attacks and stroke. The most commonly used drug for thrombolytic therapy is tissue plasminogen activator (tPA)." (Medline Plus 2014). On the other hand, "fibrinolytic therapy involves the use of drugs that dissolve clots by breaking down fibrin—a protein that connects with another sticky element of the blood known as platelets to form clots" (*Canadian Medical Association Journal 2001*). Briefly, "thrombolytics break down clots, and fibrinolytics specifically break down the fibrin in the clot" (Answers 2014).

The thrombolytic drugs, with brand names given in parentheses, include

1. Tissue plasminogen activators (tPA)
   a. Alteplase (Activase)
   b. Reteplase (Retavase)
   c. Tenecteplase (TNKase)
2. Anistreplase (Eminase)
3. Streptokinase (Kabikinase, Streptase)
4. Urokinase (Abbokinase)

"These drugs are often administered in combination with anticoagulant drugs, such as intravenous heparin or low-molecular-weight heparin, for synergistic antithrombotic effects and secondary prevention" (Wikipedia Contributors 2014b).

During the 1980s and early 1990s, four large, randomized, placebo-controlled trials showed that intravenous thrombolytic therapy with streptokinase improved survival in AMI subjects (GISSI-1 1986; ISAM 1986; ISIS-2 1988; and EMERAS 1993) (see Sections 11.2.2 and 11.3.3). However, the antigenicity and minimal fibrin specificity of streptokinase were undesirable. These undesirable properties spurred interest in finding other potential thrombolytic agents. In February 1984, a 57-year-old woman with an acute occlusion of her left anterior descending artery was the first patient who was successfully treated with recombinant tPA at the Johns Hopkins Hospital (Maroo and Topol 2004). In 1993, the Global Utilization of Streptokinase and Tissue Plasminogen Activator for Occluded Coronary Arteries (GUSTO) study showed that tPA resulted in a 15% reduction in mortality, as compared to streptokinase (GUSTO Investigators 1993) (see Section 11.2.3).

In 1992, the Cardio-Renal Advisory Committee recommended that a new thrombolytic must retain at least 50% of the benefit of the control thrombolytic. In a meta-analysis paper published by the Fibrinolytic Therapy Trialists' (FTT) Collaborative Group (1994) (see Section 11.2.2), the benefit of streptokinase over placebo was placed at 2.6% in overall mortality rates. Based on this information, reteplase was compared with streptokinase in an equivalence trial to show that it retains at least 50% of the mortality benefit (i.e., a margin of 1.3% was used). Activase and reteplase had been approved by the FDA based on their comparisons with streptokinase in equivalence studies (Gupta et al. 1999).

11.2.2 Meta-Analysis Conducted by the Fibrinolytic Therapy Trialists' Collaborative Group

The Fibrinolytic Therapy Trialists' Collaborative Group (FTT Collaborative Group, 1994) conducted a meta-analysis of the effects of treatment on mortality and on major morbidity in various patient categories in those trials designed to randomize more than 1,000 patients with suspected AMI between fibrinolytic therapy and control. There were nine such trials: GISSI-1, ISAM, AIMS, ISIS-2, ASSETS, USIM, ISIS-3 ("uncertain indication" group), EMERAS, and LATE (see Table 11.1), and the number of patients included in these studies were 11,802, 1,741, 1,254, 17,187, 5,012, 2,201, 9,158, 4,534, and 5,711, respectively.

In ISIS-3 (1992), patients for whom their physicians thought there was a "clear" indication for fibrinolytic therapy (n = 36,381) were randomized equally between SK, tPA, and APSAC, whereas those for whom the indication was considered "uncertain" (n = 9,475) were randomized equally between fibrinolytic therapy (SK, tPA, or APSAC) and open control (i.e., no placebo). On the other hand, the FTT Collaborative Group (1994) reported only 9,158 patients in the "uncertain" subgroup.

ISIS-3 (1992) reported the results among all 41,299 patients, whether in the "clear" or "uncertain" indication category, who were to receive fibrinolytic therapy. The agent was streptokinase (SK) in four; anistreplase (APSAC) in one; tissue plasminogen activator (tPA; alteplase or duteplase) in two; urokinase (UK) in one; and a random choice of SK, tPA, or APSAC in one. Six trials were placebo controlled, while three trials randomized patients

TABLE 11.1

Design Characteristics of Trials That Randomized More than 1,000 Patients between Fibrinolytic Therapy and Control

Design feature	GISSI-1	ISAM	AIMS	ISIS-2	ASSET	USIM	ISIS-3*	EMERAS	LATE
Fibrinolytic regimen									
Dose	SK, 1.5 MU	SK, 1.5 MU	APSAC, 30 U	SK, 1.5 MU	tPA 100 mg	UK, 1 MU ×2	SK, 1.5 MU; tPA, 0.6 MU/kg; or APSAC, 30 U	SK, 1.5 MU	tPA 100 mg
Duration	1 h	1 h	5 min	1 h	3 h	Bolus repeated at 60 min	1 h; 4 h; 3 min	1 h	3 h
Control	Open	Placebo	Placebo	Placebo	Placebo	Open	Open	Placebo	Placebo
Routine antiplatelet	No	Aspirin (single iv bolus)	No	Aspirin (50%)	No	No	Aspirin	Aspirin	Aspirin
Routine heparin	No	Yes, iv	Yes, iv at 6 h	No	Yes, iv	Yes, iv	50%, sc	No	64%, iv
Dose		5,000 U + 800–1,000 U/h	1,000–1,500 U/h		5,000 U + 1,000 U/h	10,000 U + 1,000 U/h	12,500 U bd		5,000 U (×1 or 2) – 1,000 U/h
Duration		72–96h, then oral anticoagulant	Until effective oral anticoagulation		24 h	48 h	7 days		48 h
Recruitment period	Jan 1984–Jul 1985	Mar 1982–Mar 1985	Sept 1985–Oct 1987	Mar 1985–Dec 1987	Nov 1986–Feb 1988	Apr 1986–Sept 1988	Sept 1989–Jan 1991	Jan 1988–Jan 1991	Apr 1989–Feb 1992

*In ISIS-3, 37,000 patients considered to have a "certain" indication for fibrinolytic therapy were randomised between SK, tPAC, and APSAC, and are not part of present report, which is restricted to those in whom indication was "uncertain." The latter were allocated half to fibrinolytic (1/3 SK, 1/3 tPA, 1/3 APSAC; all taken together in this report) and half to open control.

between fibrinolytic therapy and "open" control (i.e., both the physician and patient knew whether or not fibrinolytic therapy had been allocated).

The nine trials included 58,600 patients with suspected AMI, among whom 6,177 (10.5%) deaths occurred during the first five weeks, 564 (1.0%) strokes, and 436 (0.7%) major noncerebral bleeds during hospitalization. Among the 45,000 patients presenting with ST elevation or bundle-branch block (BBB), there was a significant absolute mortality reductions of about 30 per 1,000 for those presenting within 0–6 hours and of about 20 per 1,000 for those presenting 7–12 hours from onset. Note that according to Table 2 presented in FTT Collaborative Group report (1994; not shown here), only 41,765 patients presented with ST elevation or BBB rather than 45,000 patients as reported.

Aspirin was given routinely to all patients in four trials and to half of the patients in one trial where aspirin allocation was random. Heparin (see Section 11.2.1) was given routinely to all patients in five trials by intravenous infusion and to half of the patients in one trial where subcutaneous heparin allocation was random.

A major eligibility criterion was the time from the onset of symptoms, with four trials including only patients presenting within 6 hours (ISAM, AIMS, ASSETS, and USIM), one including only those presenting within 12 hours (GISS-1), and the other four including patients up to 24 hours (ISIS-2, ISIS-3, EMERAS, and LATE) from the onset of symptoms: in the nine trials, 22% of patients presented at 7–12 hours and 16% at 13–24 hours. The effects of the fibrinolytic agent by hours from onset are summarized in Table 11.2.

TABLE 11.2

Results of 35-Day Mortality: Fibrinolytic versus Placebo (Nine Studies)

Hours from onset	Fibrinolytic			Placebo			Relative Risk	90% Confidence Interval*	
	n	deaths	%	n	deaths	%			
0–1	1,678	159	9.48%	1,670	217	12.99%	0.7292	0.6201	0.8575
2–3	8,297	683	8.23%	8,315	889	10.69%	0.7699	0.7110	0.8338
4–6	8,294	802	9.67%	8,195	945	11.53%	0.8385	0.7782	0.9036
7–12	6,478	719	11.10%	6,404	813	12.70%	0.8743	0.8078	0.9462
13–24	4,568	457	10.00%	4,701	493	10.49%	0.9540	0.8622	1.0556
0–6	18,269	1,644	9.00%	18,180	2,051	11.28%	0.7977	0.7575	0.8399
0–24	29,315	2,820	9.62%	29,285	3,357	11.46%	0.8392	0.8065	0.8732

Source: Table 4 of FTT Collaborative Group (1994).
* See Appendix 11.A for computation of the 90% confidence interval.

11.2.3 The GUSTO Trial

Before the trial began, the FDA requested a **pilot study** of 100 patients treated with a combination of tPA and streptokinase along with intravenous heparin. After the dose regimen was found not to be associated with excessive bleeding (Granger et al. 1991), enrollment in the main GUSTO trial began on December 27, 1990, and was completed on February 22, 1993. A total of 1,081 hospitals in 15 countries in North America and Europe and in Israel, Australia, and New Zealand participated. A total of 41,021 patients with evolving MI were randomly assigned to four different thrombolytic strategies: (1) streptokinase with subcutaneous heparin (see footnote 3 to Table 11.3), (2) streptokinase with intravenous heparin, (3) accelerated tPA with intravenous heparin, or (4) a combination of streptokinase plus tPA with intravenous heparin. "Accelerated" refers to the administration of tPA over a period of one and a half hours—with two-thirds of the dose given in the first 30 minutes—rather than the conventional period of three hours. A major inclusion criterion was presenting within six hours after the onset of symptoms of MI. The primary endpoint was 30-day all-cause mortality. The results are shown in Table 11.3.

The interim analysis plan as described by GUSTO Investigators (1993, p. 675) is given in the following:

> Prespecified interim analyses of safety were performed when enrollment reached 11,274, 21,926, and 28,312 patients, with the data reviewed by an independent Data and Safety Monitoring Board. Comparisons of efficacy at the interim analyses were monitored with two-sided, symmetric O'Brien-Fleming boundaries generated with the Lan-DeMets approach to group-sequential testing (O'Brien and Fleming 1979; Lan and DeMets 1983). All tests of significance were two tailed, and treatments were compared according to the intention-to-treat principle.

TABLE 11.3

30-Day All-Cause Mortality in the GUSTO Study

Treatment	n (randomized)	n (included)	30-day Mortality Rate[1]	Deaths[2]
tPA90 and IV Heparin	10,396	10,344	6.3%	652
SK and IV Heparin	10,410	10,377	7.4%	768
SK and SC Heparin[3]	9,841	9,796	7.2%	705
tPA90 and SK and IV Heparin	10,374	10,328	7.0%	723
Total	41,021	40,845		2,848

[1] Mortality rates were reported.
[2] Numbers of deaths were calculated and were subjected to rounding-off error.
[3] This study arm was added after the first 1,160 subjects had been enrolled based on the preliminary results of ISIS-3 (ISIS-3 Collaborative Group 1993) in March 1991.
tPA90: accelerated tissue plasminogen activator; SK: streptokinase
IV: intravenous; SC: subcutaneous.

11.3 CBER Criteria in the Assessment of New Thrombotics in June 1999

11.3.1 Assessment of Thrombolytic Agent Prior to June 1999 in CBER

Prior to June 1999, CBER determined that the appropriate phase 3 study design to evaluate the safety and efficacy of a new thrombolytic agent in AMI patients required the use of a randomized design, with an established active agent as the comparator and mortality as the efficacy endpoint (FDA/CBER Memorandum 1999). The efficacy estimate should be expressed as the one-sided 95% upper confidence limit for the relative risk of mortality with the established thrombolytic agent versus mortality with placebo. The 50% of this estimate of risk reduction of the comparator is used as the noninferiority (NI) margin for comparing the new agent versus the comparator.

One basic assumption was that all thrombolytic agents are largely alike in their efficacy. Thus, use of a meta-analysis of placebo-controlled studies of thrombolytic agents, combining studies that employed different agents, is suitable. Such a meta-analysis has been conducted by the Fibrinolytic Therapy Trialists' Collaborative Group, as discussed in Section 11.2.2. Using the results of this meta-analysis, the NI margin in terms of r (see Section 3.4; adapted to the relative risk) was determined to be 1.122 for a trial of a new thrombolytic agent versus any other available thrombolytic agent (FDA/CBER Memorandum 1999). See Section 11.4 for further details. The CBER also conducted a similar meta-analysis in 1998, which is discussed in Section 11.3.2.

11.3.2 Meta-Analyses Conducted by FDA/CBER

The CBER performed comparative meta-analyses of thrombolytic agents in the treatment of AMI in early 1998 (FDA/CBER Memorandum 1998). Twenty-five studies were identified: 12 were placebo controlled [4 SK vs. PBO; 5 tPA vs. PBO; 3 APSAC or UK vs. PBO], 4 compared tPA to SK, 6 compared tPA to a thrombolytic other than SK, 2 compared tPA regimens, and 1 compared two active comparators (SK to reteplase). The relative risks of (1) streptokinase (SK) over placebo (PBO) and (2) alteplase (tPA) over placebo were estimated, focusing on treatment within 6 and 24 hours of onset (see Table 11.4). In addition, the relative risks of two active agents in several pairwise comparisons of agents or regimens were estimated, focusing on treatment within 6 hours of onset (see Table 11.5). These estimates were then combined to give a variety of estimates of tPA effect versus placebo (see Table 11.6). Several principles were combined

TABLE 11.4

Relative Risk of SK and tPA versus Placebo

Estimated of Thromboltic Effect Based on Direct Comparison to Placebo				
Comparsion	Notes	#Studies, #Patients	Relative Risk	90% C.I.
SK-6° vs. PBO	Only data known to be 0–6 hr from onset	4, 22,857	0.77	0.72, 0.83
SK-24° vs. PBO	Data from up to 24 hr from onset	4, 35,174	0.82	0.78, 0.86
tPA-6° vs. PBO	Only 0–6 hr from onset; all using 3hr regimen of tPA	3, 5,875	0.73	0.63, 0.85
tPA-24° vs. PBO	Up to 24 hr from onset	5, 11,758	0.80	0.72, 0.89
Any Thrombolytic vs. PBO (FTT Analysis)	Only 0–6 hr from onset; only ST Elev or BBB	8, 28,824	0.76	0.72, 0.80

Source: FDA/CBER Memorandum (1998).

TABLE 11.5

Relative Risk of tPA Regimens versus SK and Other Agents

Estimated of Thromboltic Effects				
Comparsion	Notes	#Studies, #Patients	Relative Risk	90% C.I.
tPA90 vs. SK	0–6 hr from onset GUSTO-1 study	1, 30,517	0.86	0.79, 0.94
tPA vs. SK	0–6 hr from onset; any tPA regimen; not including duteplase (ISIS-3)	3, 51,285	0.95	0.90, 1.00
tPA90 vs. rPA	Moslty 0–6 hr from onset	2, 15,413	0.98	0.88, 1.08
rPA vs. SK	Up to 12 hr from onset	1, 5,990	0.95	0.82, 1.09

Source: FDA/CBER Memorandum (1998).

in various ways to derive each of the estimates. Assumptions include whether or not

1. tPA (as an agent; irrespective of regimen) is intrinsically different from other agents.
2. The tPA 90-minute regimen is different from the 3-hour regimen.
3. Indirect (chained) comparisons can carry weight.

TABLE 11.6

Relative Risk of tPA Regime versus Placebo: Direct and Indirect Estimates

	Estimates of tPA Effect Compared to Placebo				
Comparison	Postulates	Notes	# Studies, # Patients	Relative Risk	90% C.I.
Any Thrombolytic vs. PBO	tPA is much the same as any other thrombolytic (FTT Analysis)	only 0–6 hr from onset; only ST Elev or BBB	8, 28,824	0.76	0.72, 0.80
tPA vs. PBO	tPA is different from other agents; only direct comparison strong	Only 0–6 hr from onset; all using 3hr regimen of tPA	3, 5,875	0.73	0.63, 0.85
tPA vs. SK • SK vs. PBO	tPA is different from other agents; indirect comparison method	0–6 hr from onset; any tPA regimen; not including duteplase (ISIS-3) only data known to be for 0-6 hr from onset	3, 51,285 3, 22,857	0.74	0.68, 0.80
tPA90 vs. SK • SK vs. PBO	tPA is different from other agents; tPA 90 minute regimen is different from 3hr regimen; indirect comparison method	0–6 hr from onset GUSTO-1 study only data known to be for 0–6 hr from onset	1, 30,517 4, 22,857	0.67	0.60, 0.75
tPA90 vs. rPA • rPA vs. SK • SK vs. PBO	tPA is different from other agents; (only tPA 90-minute regi men available to compare to rPA); indirect comparison method	Mostly 0–6 hr from onset; some data up to 12 hr from onset included for rPA vs. SK; only data known to be for 0–6 hr from onset for SK vs. PBO	2, 15,413 1, 5,990 4, 22,857	0.72	0.59, 0.86

Source: FDA/CBER Memorandum (1998).

Note that the results of "any thrombolytic vs. PBO" in Tables 11.4 and 11.6 (1) include patients from only eight studies who were treated within 6 hours of onset and had clear electrocardiograph (EKG) changes of ST segment elevation or bundle-branch block (ST Elev or BBB) and (2) does not include the placebo-controlled stratum of ISIS-3 (1992) (see Section 11.2.2).

11.3.3 Relative Risk of tPA90 versus Placebo

In the mid-1990s, there were two phase 3 studies where the 90 minute regimen of tPA90 was used as the comparator treatment. These studies limited subject enrollment to patients within 6 hours of onset of chest pain, because a lessening of efficacy of thrombolytic therapy with longer time since onset, particularly beyond the 6-hour eligibility limit, is well established. Issues had been raised specifically regarding the comparison to tPA90 as the active control. Presumably, the issues were related to the assumption that all thrombolytic agents are largely alike in their efficacy. Without such an assumption, the NI margin has to be determined for a specific regimen of a specific thrombolytic agent to be used as an active control in the trials.

There are no adequate placebo-controlled studies to directly estimate the relative risk of tPA90 versus placebo. However, this relative risk, denoted by tPA90/placebo, may be expressed as

$$\text{tPA90/placebo} = (\text{tPA90/SK}) \times (\text{SK/placebo}) \tag{11.1}$$

where tPA90/SK and SK/placebo denote the relative risk of (1) tPA90 versus SK and (2) SK versus placebo, respectively. tPA90/SK may be estimated from the GUSTO study (see Section 11.2.3), and SK/placebo may be estimated from four placebo-controlled studies of streptokinase (see Sections 11.2.2 and 11.3.2) in two ways: (1) simple pooling and (2) using Poisson regression. Only data pertaining to patients enrolled within 6 hours of onset were incorporated into the calculations. The results are summarized in Table 11.7.

Although GUSTO is an open label study, it was reviewed in detail by CBER and found to be acceptable for estimating tPA90/SK. The FDA determined that a more conservative approach should be applied to use of estimations from the single GUSTO trial than from the multiple and consistent SK versus placebo trials by using a 99.5% confidence interval (presumably one-sided) rather than a 95% confidence interval. This approach is equivalent to weighting the GUSTO findings by a factor determined as $z_{0.95}/z_{0.995}$ = 0.6386, where $z_{0.95}$ = 1.64485 and $z_{0.995}$ = 2.57583. However, it is not clear why 0.6372 instead of 0.6386 was used in FDA/CBER Memorandum (1999), although the impact of such a difference is minimal.

Combining the results of the GUSTO study and four studies comparing SK with placebo gives an estimate of the relative risk of tPA90 versus placebo. There are two ways of estimating SK/placebo (simple pooling or Poisson regression), and the GUSTO study can be unweighted or weighted, resulting in four possible estimates of tPA90/placebo. These results, with the 90% confidence intervals, are summarized in Table 11.8.

TABLE 11.7

All-Cause Mortality of Five Studies (Within 6 Hours from Onset of Symptoms)

	Agent 1			Agent 2			Relative Risk	90% Confidence Interval
	n	Deaths	%	n	Deaths	%		
tPA90 vs. SK								
GUSTO-1 (1993)[2]	10,344	652	6.30%	20,173[1]	1,473	7.30%	0.8632	(0.8010, 0.9302)*
SK vs. Placebo[3]								
GISSI (1986)[4]	4,865	495	10.17%	4,878	623	12.77%	0.7967	
ISAM (1986)[4]	859	54	6.29%	882	63	7.14%	0.8801	
ISIS-2 (1988)[5]	5,350	471	8.80%	5,360	648	12.09%	0.7282	
EMERAS (1993)[5,6]	336	47	13.99%	327	47	14.37%	0.9732	
Simple pooling	11,410	1,067	9.35%	11,447	1,381	12.06%	0.7751	(0.7275, 0.8259)*
Poisson regression							0.7744	(0.7242, 0.8281)**

[1] Combined (1) SK and IV Heparin and (2) SK and SC Heparin from Table 11.3.
[2] Primary endpoint: 30-day all-cause mortality.
[3] Using the same data as in Table 11.4 for "SK-6° vs. PBO."
[4] Primary endpoint: 21-day all-cause mortality.
[5] Primary endpoint: 35-day all-cause mortality.
[6] Deaths in hospital rather than deaths in days 0–35.
* See Appendix 11.A for computation of the 90% confidence interval.
** Calculated using SAS PROC GENMOD (see Appendix 11.B).

TABLE 11.8

Relative Risk of tPA90 versus Placebo (Within 6 Hours after Onset of Symptoms)

SK vs. Placebo	GUSTO	Relative Risk	90% Confidence Interval[1]
Simple pooling	Unweighted	0.6691	(0.6066, 0.7380)
	Weighted	0.6691	(0.5976, 0.7492)
Poisson regression[2]	Unweighted	0.6685	(0.6046, 0.7391)
	Weighted[2]	0.6685	(0.5958, 0.7501)

[1] See Appendix 11.C for the computation of the 90% confidence interval.
[2] Used in FDA/CBER Memorandum (1999). See the fifth row of Table 11.6.

11.3.4 Calculations of the NI Margin r

The upper limit of the 90% confidence interval for tPA90/placebo is calculated as 0.7777 and is used as the maximum expected relative risk of tPA90 versus placebo. Applying the 50% rule of required retained efficacy, the maximum acceptable relative risk for new agent versus placebo is 0.8888. The ratio of these two numbers (0.8888/0.7777 = 1.1429) directly leads to the calculation of the maximum acceptable relative risk of new agent versus tPA90 (FDA/CBER Memorandum 1999). That is r = 1.1429 (see Section 3.4).

It is not clear why 0.7777 rather than 0.7501 (see Table 11.8) was used. To understand why r is given by the ratio of 0.8888 over 0.7777, let us assume that the true relative risk of tPA90 over placebo is equal to 0.7777. That is, tPA90/placebo = 0.7777. The percent risk reduction of tPA90 over placebo is given by (1–tPA90/placebo) = 0.2223. The 50% rule of required retained efficacy leads to maximum risk allowed for the new thrombolytic relative to placebo given by

$$tPA90/placebo + 0.5 \, (1 - tPA90/placebo) = 0.7777 + 0.5 \cdot 0.2223 = 0.8888$$

The maximum risk allowed for the new thrombolytic relative to tPA90 is then calculated as

$$= 0.8888/0.7777 = 1.1429$$

Using the notations given in Section 1.7 of Chapter 1 and the hypotheses formulated in Section 3.4 of Chapter 3 (adapted to binary data), that is, S = tPA90 and P = Placebo, then

$$r = [S/P + 0.5(1 - S/P)]/(S/P) = [S + 0.5(P - S)]/S$$

In the previous formula, the 50% retention is applied to the effect of S on the original scale, that is, (P − S). It is then converted to the risk relative to S. This is in contrast to the proposed r given by Equation 3.3 in Section 3.4 of Chapter 3, where the 50% retention is applied to the effect of S on the log scale. Using this approach, we have $r = 1/0.7777^{0.5} = 1.1340$.

11.3.5 Limitations in the Determination of r

The determination of r in Section 11.3.4 has the following limitations:

1. Although all-cause mortality is the primary endpoint for all studies, the time used in the determination of the mortality rates varies from study to study. See the footnotes to Table 11.7.

2. Although the analyses were restricted to patients who were treated within 6 hours from onset of symptoms, randomization was presumably not stratified by hours from onset. GUSTO-1 (1993) and ISAM (1986) enrolled patients who presented at the hospitals up to

6 hours, GISSI (1986) up to 12 hours, and ISIS-2 (1988) and EMERAS (1993) up to 24 hours. Thus, randomization in the latter three studies is less than "optimal."

3. Since the 30-day mortality rate reported in the GUSTO study is rounded to three decimal places, back-calculated "Deaths" in Table 11.3 are within ±5.

4. Since r is determined based on an indirect comparison of tPA90 with a placebo, the constancy assumption (see Section 2.5.1 of Chapter 2) is of a greater concern than if it were based on a direct comparison.

11.4 Discussion

Since the relative risk is the metric for comparison, the 50% rule of required retained efficacy should be applied on a log scale rather than the original scale (see Section 11.3.4). Furthermore, we need to assess the constancy assumption because 50% is used as the retention of the control effect (see Section 2.4 of Chapter 2 and Section 5.7 of Chapter 5).

Based on the results from the FTT Collaborative Group (1994), the upper limit of the 90% confidence interval for the relative risk (RR) of any thrombolytic (given within 6 hours from onset) over placebo is 0.8399 (see Table 11.2 in Section 11.2.2), and r is determined to be 1.0953 [= (0.8399 + 0.5·0.1601)/0.83899], with 50% retention being calculated on the original scale. On the other hand, based on the FDA/CBER Memorandum (1998), the upper limit of the 90% confidence interval for the RR of any thrombolytic given within 6 hours from onset over placebo is 0.80 (see "any Thrombolytic vs. PBO" in Tables 11.4 and 11.6 in Section 11.3.2), and r is determined to be 1.125 [= (0.80 + 0.5·0.20)/0.80], which is within the rounding-off error from 1.122. It appears that r = 1.122 is obtained based on the results from the FDA/CBER Memorandum (1998) rather than the FTT Collaborative Group (1994).

The FTT Collaborative Group (1994) noted that the "uncertain" cohort (n = 9475) in the ISIS-3 study (see Section 11.2.2) included a large proportion of patients without ST elevation or BBB. Therefore, the thrombolytic effect observed in this cohort (n = 9475) in the ISIS-3 study is most likely to be less than what is expected in the target population.

Prespecified interim analyses were performed in the GUSTO study (see Section 11.2.3). It is not clear how such interim analyses should be taken into account in estimating tPA90/placebo.

In Section 5.3 of Chapter 5, the lower limit rather than the upper limit of the confidence interval is used because a large value corresponds to a better outcome. Furthermore, a 95% confidence interval rather than a 90% confidence interval is used, leading to a smaller effect size, and hence a tighter

NI margin, than using the 90% confidence interval. FDA draft guidance (2010) suggests using the 95% confidence interval in the examples given in the appendix to the document.

There are other alternatives to estimate SK/placebo in addition to the methods used in Table 11.7 in Section 11.3.3. See, for example, McNutt et al. (2003). However, a comparison of these methods and a recommendation of what method to use are beyond the scope of this book.

Appendix 11.A Confidence Interval for the Relative Risk

Suppose that X_n and Y_m are independent and follow a binomial distribution with parameters (n, p) and (m, q), respectively, and $RR = p/q$ denotes the relative risk. Let $\hat{p} = X_n/n$, $\hat{q} = X_m/m$, and $RR = \hat{p}/\hat{q}$ denote the estimates for p, q, and RR, respectively. Variance of $\log_e(\hat{p})$ is approximately equal to $(1-p)/(np)$ and variance of $\log_e(\hat{q})$ is approximately equal to $(1-q)/(mq)$ by using the delta method (Wikipedia Contributors 2014c), that is,

$$\mathrm{Var}[\log_e(\hat{p})] \approx (1-p)/(np)$$

and

$$\mathrm{Var}[\log_e(\hat{q})] \approx (1-q)/(mq)$$

Therefore,

$$\mathrm{Var}[\log_e(RR)] \approx (1-p)/(np) + (1-q)/(mq)$$

which can be estimated by

$$(1-\hat{p})/(n\hat{p}) + (1-\hat{q})/(m\hat{q}) \tag{11.2}$$

and the 90% confidence interval for RR on the log scale is calculated as

$$\log_e(\hat{p}) - \log_e(\hat{q}) \pm 1.645\sqrt{(1-\hat{p})/(n\hat{p}) + (1-\hat{p})/(m\hat{q})}$$

Exponentiating this confidence interval gives the 90% confidence interval for RR.

Appendix 11.B Poisson Regression Implemented by SAS PROC GENMOD

This appendix presents the SAS codes and the results of the Poisson regression analysis shown in Table 11.7.

```
data A;
        input n y TRT$ study;
        ln = log(n);
        datalines;
4865   495   SK        1
4878   623   Placebo   1
859    54    SK        2
882    63    Placebo   2
5350   471   SK        3
5360   648   Placebo   3
336    47    SK        4
327    47    Placebo   4
        ;
        run;

proc genmod data = A;
        class TRT Study;
        model y = TRT study/dist= poisson
                link= log
                offset = ln;
        estimate "beta" TRT -1 1/exp alpha = 0.1;
run;
```

		Contrast Estimate Results								
	Mean	Mean		L'Beta	Standard		L'Beta		Chi-	Pr >
Label	Esitmate	Confidence Limits		Estimate	Error	Alpha	Confidence Limits		Square	ChiSq
beta	0.7744	0.7242	0.8281	-0.2557	0.0408	0.1	-0.3227	-0.1886	39.35	<.0001
Exp (beta)				0.7744	0.0316	0.1	0.7242	0.8281		

Appendix 11.C Confidence Interval for the Relative Risk Using Chain Studies

As discussed in Section 11.3.3, there is no direct comparison of tPA90 with placebo, and the RR of tPA90 over placebo may be expressed as the products of two RRs comparing (1) tPA90 versus SK and (2) SK versus placebo by chain studies, as shown in Equation 11.1. Let RR = tPA90/placebo, RR_1 = tPA90/SK, and RR_2 = SK/placebo. Then RR = $RR_1 \cdot RR_2$ and

$$\text{Var}[\log_e(RR)] = \text{Var}[\log_e(RR_1)] + \text{Var}[\log_e(RR_2)]$$

$\text{Var}[\log_e(RR_1)]$ and $\text{Var}[\log_e(RR_2)]$ can be estimated by Equation 11.2, as shown in Appendix 11.A. With these estimates, the 90% confidence interval for $\log_e(RR)$ may be calculated as

$$\log_e(\widehat{RR}_1) + \log_e(\widehat{RR}_2) \pm 1.645 \sqrt{\text{Var}[\log_e(\widehat{RR}_1)] + \text{Var}[\log_e(\widehat{RR}_2)]}$$

Exponentiating this confidence interval gives the 90% confidence interval for RR. For the weighted estimate,

$$\text{Var}[\log_e(RR)] = \text{Var}[\log_e(RR_1)] + \text{Var}[\log_e(RR_2)]/0.6386$$

References

Canadian Medical Association Journal (CMAJ), "Questions and answers on fibrinolytic (clot-dissolving) therapy for acute myocardial infarction (heart attack)," September 18, 2001; 165 (6), pp. 796-797. http://www.collectionscanada.gc.ca /eppp-archive/100/201/300/cdn_medical_association/cmaj/vol-165/issue-6 /pdf/pg791app1.pdf (Accessed January 12, 2014).

EMERAS (Estudio Multicéntrico Estreptoquinasa Repúblicas de América del Sur) Collaborative Group (1993). Randomised Trial of Late Thrombolysis in Patients with Suspected Acute Myocardial Infarction. *Lancet*, **342**:767–72.

FTT Collaborative Group (1994). Indications for Fibrinolytic Therapy in Suspected Acute Myocardial Infarction: Collaborative Overview of Early Mortality and Major Morbidity Results from all Randomized Trials of More than 1000 patients. *Lancet*, **343**:311–322.

GISSI (Gruppo Italiano per lo Studio della Streptochinasi nell'Infarto miocardico)(1986). Effectiveness of Intravenous Thrombolytic Treatment in Acute Myocardial Infarction. *Lancet*, **327**:397–402.

Granger C, Califf R, Woodlief L, et al. (1991). The Global Utilization of Streptokinase and Tissue Plasminogen Activator for Occluded Coronary Arteries (GUSTO) Pilot Study: Combined Streptokinase and t-PA. *Circulation*, **84**:Suppl II:II–573 abstract.

Gupta G, Hsu H, Ng T-H, Tiwari J, and Wang C (1999). Statistical Review Experiences in Equivalence Testing at FDA/CBER. *Proceedings of the Biopharmaceutical Section, American Statistical Association*, 220–223.

GUSTO Investigators (1993). An International Randomized Trial Comparing Four Thrombolytic Strategies for Acute Myocardial Infarction. *New England Journal of Medicine*, **329**:673–682.

ISAM (Intravenous Streptokinase in Acute Myocardial Infarction) Study Group (1986). A Prospective Trial of Intravenous Streptokinase in Acute Myocardial Infarction (ISAM): Mortality, Morbidity, and Infarct Size at 21 Days. *New England Journal of Medicine*, **314**:1465–1471.

ISIS-2 (Second International Study of Infarct Survival) Collaborative Group (1988). Randomised Trial of Intravenous Streptokinase, Oral Aspirin, Both, or Neither Among 17, 187 Cases of Suspected Acute Myocardial Infarction: ISIS-2. *Lancet*, **332**:349–60.

ISIS-3 (Third International Study of Infarct Survival) Collaborative Group (1992). ISIS-3: A Randomised Trial of Streptokinase vs. Tissue Plasminogen Activator vs. Anistreplase and of Aspirin Plus Heparin vs. Aspirin Alone Among 41,299 Cases of Suspected Acute Myocardial Infarction. *Lancet*, **339**:753–770.

Lan KKG and DeMets DL (1983). Discrete Sequential Boundaries for Clinical Trials. *Biometrika*, **70**:659–663.

Maroo A and Topol EJ (2004). The Early History and Development of Thrombolysis in Acute Myocardial Infarction. *Journal of Thrombosis and Haemostasis*, **2**:1867–1870.

McNutt L-A, Wu C, Xue X, and Hafner JP (2003). Estimating the Relative Risk in Cohort Studies and Clinical Trials of Common Outcomes. *American Journal of Epidemiology*, **157**:940–943.

Medline Plus (2014). Thrombolytic Therapy. http://www.nlm.nih.gov/medline plus/ency/article/007089.htm (Accessed: January 11, 2014).

O'Brien PC and Fleming TR (1979). A Multiple Testing Procedure for Clinical Trials. *Biometrics*, **35**:549–556.

U.S. FDA/CBER Memorandum (1998). Thrombolytic Agents in the Treatment of Acute MI: CBER Comparative Meta-Analyses. April 1998.

U.S. FDA/CBER Memorandum (1999). Summary of CBER Considerations on Selected Aspects of Active Controlled Trial Design and Analysis for the Evaluation of Thrombolytics in Acute MI. June 1999.

U.S. FDA (2010). Draft Guidance for Industry: Non-Inferiority Clinical Trials. http://www.fda.gov/downloads/Drugs/GuidanceComplianceRegulatoryInformation/Guidances/UCM202140.pdf (Accessed: August 25, 2013).

"What is the difference between thrombolytic agents and fibrinolytic agents and anticoagulant agents?" (2014). Answers. http://www.answers.com/Q/What_is _the_difference_between_thrombolytic_agents_and_fibrinolytic_agents_and_ anticoagulant_agents (Accessed January 12, 2014).

Wikipedia Contributors (2014a). Myocardial Infarction. http://en.wikipedia.org /wiki/Myocardial_infarction (Accessed: January 16, 2014).

Wikipedia Contributors (2014b). Thrombolytic Drug. http://en.wikipedia.org/wiki /Thrombolytic_drug (Accessed: January 19, 2014).

Wikipedia Contributors (2014c). Delta Method. http://en.wikipedia.org/wiki /Delta_method (Accessed: February 13, 2014).

12

Issues and Challenges

12.1 Introduction

The null hypothesis of a specified difference had been formulated as early as the 1970s (e.g., Remington and Schork 1970; Dunnett and Gent 1977; Makuch and Simon 1978). Mathematically, it appears to be not much different from testing the null hypothesis of no difference because the t-statistic can be calculated by subtracting the specified difference from the difference in the sample means. In reality, however, it is not as simple as it looks. For noninferiority (NI) trials, the specified difference is known as the NI margin. In Chapter 2, the proposed NI margin as a small fraction of the therapeutic effect of the active control in a clinical trial is conceptually very simple. In practice, however, there are many issues in the determination of the NI margin and in the analysis of an NI trial as seen throughout this book. This chapter covers many other issues and challenges that need to be addressed, although many advances have been made over the past three decades.

A search on the PubMed website for articles with "noninferiority" in title/abstract was performed. The number of publications by year of publication is shown in Figure 12.1. In spite of the limitations of such a search, this figure shows how much interest in this research area has grown during the last decade.

Five special issues on noninferiority were published by three journals in the mid-2000s: two by *Statistics in Medicine* (2003, 2006), two by the *Journal of Biopharmaceutical Statistics* (2004 and 2007), and one by the *Biometrical Journal* (2005). One additional special issue on NI trials was published by *Biopharmaceutical Statistics* in 2011. In addition, there were two workshops on NI trials in the 2000s: PhRMA (Pharmaceutical Research and Manufacturers of America) Non-Inferiority Workshop in 2002 and FDA-Industry Workshop in 2007. Finally, a short course on NI trials was offered at the FDA-Industry Workshop in 2003. These tremendous research efforts in NI trials in the past 15 years have made many advances in this area. However, many issues remain unresolved.

This chapter highlights (1) the fundamental issues (see Section 12.2), (2) advances (see Section 12.3), (3) current controversial issues (see Section 12.4), and (4) issues and challenges (Section 12.5) in the design and analysis of NI trials.

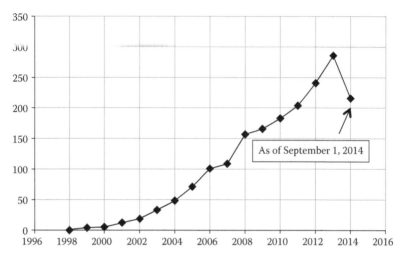

FIGURE 12.1
Number of publications by year.

12.2 Fundamental Issues

Many issues in NI trials are not seen in superiority trials, although the test statistics (e.g., t-test) used the NI, and superiority trials do not "look" much different mathematically (see Section 12.1). The three fundamental issues are (1) lack of internal validation of assay sensitivity (see Sections 2.6 and 2.7), (2) need for a strong, unverifiable constancy assumption (see Sections 2.5.1 and 2.7), and (3) bias toward equality.

To validate assay sensitivity, one has to rely on the historical placebo-controlled trials of the active control, which are, very often, inadequate. Although discounting (see Section 2.5.2 of Chapter 2) may alleviate the concern with constancy assumption, such discounting is subjective, and there is no scientific basis to determine the discounting factor. It is extremely difficult, if not impossible, to assess the publication bias in the meta-analysis in the estimation of the control effect. Under the null hypothesis of a specific difference, poorly conducted studies (e.g., mixing up treatment assignment) would bias toward equality. This would potentially lead to (1) a fault rejection of the null hypothesis and (2) an increase in the Type I error rate although the variability may be increased. Unlike the superiority trial, such a bias may happen without unblinding.

In the two-arm NI trials, the historical data is used to either (1) come out with the NI margin using the fixed margin method (Section 5.3) or (2) be incorporated the into the test statistic using the synthesis approach (Section 5.4). The most fundamental issue in the analysis of NI trials is that patients were not "randomized" into the two studies—the current NI trial and the "super-study" of the historical studies. We have to rely on the constancy assumption or discounting in the assessment of efficacy, as compared to placebo or percent preservation. Therefore, efficacy assessment in NI trials is less credible than that in the placebo-controlled trial (or superiority trial) or the three-arm, gold-standard design (see Section 8.2 of Chapter 8). On the other hand, the constancy assumption is weaker than the nonrandomized, historical-controlled studies, where the historical control mean is assumed to be the same as in the current trial (see Section 2.5.1 of Chapter 2).

12.3 Advances in Noninferiority Trials

Design and analysis of an active-control trial have gone through two major transitions over the past four decades. The first transition dealt with the formulation of the hypotheses, and the second transition dealt with the study objective. These transitions happened very slowly.

12.3.1 Formulation of Noninferiority Hypothesis and Noninferiority Margin

Although the null hypothesis of a specified difference was suggested in the 1970s (see Section 12.1), based on the lists in Sections 1.3 and 1.4 of Chapter 1, the formulation of the hypotheses given by Equation 1.3 in Section 1.7 of Chapter 1 was widely recognized and accepted two decades later in the 1990s. With such a formulation of the hypotheses, researchers were facing a great challenge in determining the NI margin (or equivalence margin, as it was called at the time; see Section 1.3 of Chapter 1). A small fraction of the effect size as the NI margin was first proposed by Ng (1993) (see Section 2.2 of Chapter 2). Such a proposal was "translated" into preservation or retention in the literature (see Section 2.4 of Chapter 2) and was used in the late 1990s. See the thrombolytic example discussed in Chapter 11.

12.3.2 Study Objective: Showing Efficacy

The hypotheses in Section 1.7 of Chapter 1 were first formulated with the objective of showing NI or equivalence in a broad sense rather than a strict sense (see Sections 1.2 and 1.6 of Chapter 1). Very often, such an objective is not practical due to the limitation of the sample size. A viable alternative is to show the efficacy of T by an indirect comparison with P. Such an objective was first discussed by Ng (1993) in determining the sample size in active-controlled trials through specification of the treatment difference. The concept of showing efficacy dates back to 1985 (see Section 5.6 of

Chapter 5). This latter objective of showing efficacy became noticeable in the literature in the 2000s (e.g., Hassalblad and Kong 2001; Wang and Hung 2003; EMEA/CPMP 2005, FDA 2010). However, showing efficacy in NI trials might not the sufficient for drug approval (FDA 2010), leading to a controversial issue (see Section 12.4.1). Furthermore, Snapinn and Jiang (2014) question the formulation of the NI hypothesis given by Equation 1.3 in Section 1.7 of Chapter 1 when the study objective is to show efficacy (see Section 12.4.2).

12.3.3 From the Fixed-Margin Method to the Synthesis Method

When the NI hypotheses given in Section 1.7 of Chapter 1 were first formulated, the NI margin δ was considered a fixed constant. Therefore, it is natural to use the fixed-margin method (see Section 5.3 of Chapter 5). However, with the NI margin depending on the effect size as given by Equation 2.1 in Chapter 2, an alternative to the fixed-margin method—namely, the synthesis method (see Section 5.4 of Chapter 5)—emerged in the 2000s. See, for example, Ng (2001); Wang and Hung (2003); Snapinn and Jiang (2008b); and Hung, Wang, and O'Neill (2009).

The concept of the synthesis method to show efficacy of the test treatment as compared to placebo dates back to 1985 (see Section 5.6 of Chapter 5). See also Hassalblad and Kong (2001) and the references therein. Based on a search of the PubMed website, it appears that the term "synthesis method" was first introduced in the mid-2000s by Wang and Hung (2003).

12.3.4 Beyond Two Treatment Groups: Gold-Standard Design

The analyses of clinical trials with more than two treatment groups are much more complicated than those of studies with just two treatment groups because there are many pairwise comparisons to be considered. Chapter 8 focused the discussion on three treatment groups. Of particular interest is the gold-standard design (STP) (see Section 8.2 of Chapter 8). The traditional strategy is to simultaneously test the three null hypotheses H_{01}, H_{02}, and H_{03} given in Section 8.2.2 of Chapter 8 at the prespecified significance level α, and the trial will be considered successful only if all three null hypotheses are rejected. The modified Koch-Rohmel procedure proposed by Rohmel and Pigeot (2011) is a step-down procedure to control the family-wise error rate (see Section 8.2.2 of Chapter 8). Briefly, this procedure first tests H_{01} at the prespecified significance level α, and if H_{01} is rejected, then tests H_{02} and H_{03} simultaneously at the same α significance level. Obviously, this is a great improvement over the traditional strategy. Note that the NI margin δ in H_{02} is considered a fixed constant. With the NI margin given by Equation 2.1 of Chapter 2, the linearized approach is a viable alternative (see Section 8.2.3 of Chapter 8).

12.3.5 Toward One Primary Analysis in Noninferiority Trials: Intention-to-Treat versus Per-Protocol

It was widely recognized in the 1990s and 2000s that the intention-to-treat (ITT) analysis was anticonservative in NI trials, while the exclusion of non-compliant subjects in the per-protocol (PP) analysis could undermine the prognostic balance between the two treatment arms achieved through randomization (see Section 10.5.2 of Chapter 10). Therefore, both analyses are currently required by regulatory agencies (see Section 10.5.1 of Chapter 10). However, recent literature has tilted the balance in favor of the ITT analysis, and such a movement is supported by the facts that the ITT analysis (1) preserves the value of randomization and (2) estimates real-world effectiveness (see Section 10.5.5 of Chapter 10).

The major hurdle in ITT analysis is the missing data. With (1) the 18 recommendations by the National Research Council (2010) (see Appendix 10.A in Chapter 10) and (2) the recent 10 mandatory standards in the prevention and handling of missing data in patient-centered outcomes research recommended by Li et al. (2014) (see Appendix 10.B of Chapter 10), the amount of missing data hopefully will be kept to minimum and the study will be of high quality so that the ITT analysis will be widely accepted as the primary analysis in NI trials.

12.4 Current Controversial Issues

12.4.1 One Standard for Approval: Efficacy versus Preservation of Control Effect

As noted in Chapter 11, preservation of the 50% control effect was used in studying thrombolytics in the late 1990s. Since then, 50% preservation is often used in the literature (e.g., Rothmann et al. 2003; Wang and Hung 2003; FDA 2004; Sorbello, Komo, and Valappil 2010). In fact, the Food and Drug Administration (FDA) draft guidance on NI (FDA 2010) states that "a typical value for M_2 is often 50% of M_1, ..." and that "choosing M_2 as 50% of M_1 has become usual practice for cardiovascular (CV) outcome studies, whereas ..." However, there has been push-back from the pharmaceutical industry in the regulatory approval of the requirement of showing greater than 50% preservation (e.g., Snapinn and Jiang 2008a; Peterson et al. 2010a; Huitfeldt and Hummel 2011; Snapinn and Jiang 2014).

Peterson et al. (2010a) argued that regardless of whether a placebo-controlled or active-control trial has been performed, the same standard of evidence should be applied. The same standard of evidence refers to showing superiority to placebo (i.e., efficacy). Showing greater than 50% preservation is clearly a higher standard than showing superiority to placebo, as illustrated in Figure 2.5b in Chapter 2. On the other hand, due to untestable assumptions in the indirect comparison with placebo in an active-control trial, Hung and Wang (2010) argued that such an indirect comparison is different from the direct comparison with placebo in a placebo-controlled trial.

Since preservation and discounting are indistinguishable mathematically (see Section 5.7 of Chapter 5), 50% preservation is subjected to two different interpretations (Snapinn and Jiang 2008a). If the study objective is to show efficacy as compared to placebo, then the preservation may be "interpreted" as the discounting. For example, Wang and Hung (2003, 153) stated the following:

> However, given possible uncertainty on cross-trial inference, in order to be fairly certain that the new drug would have been superior to placebo had the placebo treatment been studied in the trial, it was decided that the new drug must be shown to preserve at least 50% of the control effect in this target population of the active-controlled trial.

In that case, "discounting" should have been used rather than "preservation." In other words, $\gamma = 0.5$ and $\varepsilon = 1$ should be used instead of $\gamma = 1$ and $\varepsilon = 0.5$, although the results are exactly the same. Equivalently, imposing some degree of conservativeness can also be done through some kind of discounting in the determination of M_1 (Huitfeldt and Hummel 2011).

On the other hand, if the study objective is to show greater than 50% preservation, then the constancy assumption is needed. A much higher preservation is probably needed for a NI claim. In any case, we should keep in mind the fundamental issues of the indirect comparison, as discussed in Section 12.2.

12.4.2 Efficacy Hypotheses in Noninferiority Trials

To be consistent with the study objective of showing efficacy, Snapinn and Jiang (2014) suggest that the null hypothesis $T \leq P$ be tested against the alternative hypothesis $T > P$, instead of the hypotheses defined by Equation 1.3 in Chapter 1. This makes sense logically at first glance. However, since there is no placebo arm in the current NI trial, a direct comparison to placebo cannot be made and formulation of such a hypothesis is questionable. A viable alternative is to make an indirect comparison to placebo through comparison to the standard therapy or active control in the current NI trial (Julious 2011; Snapinn and Jiang 2011). The formulation of the hypotheses for such an indirect comparison is given by Equation 5.1 in Chapter 5, with $\varepsilon = 1$ for efficacy, where $\gamma = 1$ if the constancy assumption holds. There are two approaches for testing the hypothesis given by Equation 5.1a, as discussed in Chapter 5, and no consensus has been reached as to which method should be used (Section 12.4.3).

12.4.3 Fixed-Margin versus Synthesis Methods

The FDA draft guidance (FDA 2010) suggests the fixed-margin method be used for showing efficacy, as it states:

> We believe the fixed-margin approach is preferable for ensuring that the test drug has an effect greater than placebo (i.e., the NI margin M_1 is ruled out). However, the synthesis approach, appropriately conducted, can be considered in ruling out the clinical margin M_2.

On the other hand, many authors (e.g., Snapinn and Jiang 2008a; Peterson et al. 2010a; Huitfeldt and Hummel 2011; Snapinn and Jiang 2014) advocate the synthesis method. For example, Huitfeldt and Hummel (2011) state:

> We propose that the most efficient method should be used for both analyses, that is, the synthesis method.

The choice of statistical approach (fixed-margin or synthesis) in the analysis of NI trials remains controversial. See, for example, Hung and Wang, (2010) and Peterson et al. (2010b). As discussed in Section 5.1 of Chapter 5, the fixed-margin method (Section 5.3) is conditioned on the historical data, while the synthesis method (Section 5.4) is unconditioned on the historical data. In testing the preservation hypotheses given by Equation 5.1 in Section 5.1, assuming the effect size is appropriately discounted, the fixed margin cannot control the unconditional Type I error rate at the α significance level exactly. More specifically, the unconditional Type I error rate will be inflated (deflected) if the effect size is overestimated (underestimated) by the one-sided $(1 - \alpha^*/2)100\%$ lower confidence limit as discussed in Section 5.8 of Chapter 5. On the other hand, the synthesis method can control the unconditional Type I error rate at the nominal $\alpha/2$ level. This is an advantage of the synthesis method. However, the results of the analysis using this method may be difficult to interpret in the sense that the clinical significance cannot be assessed (Hung, Wang, and O'Neil 2007).

As discussed in Section 5.5 of Chapter 5, from a statistical point of view, the synthesis method rather than the fixed-margin method should be used; however, from a practical point of view, the fixed-margin method rather than the synthesis method should be used.

12.5 Issues and Challenges

The most fundamental issue in the two-arm NI trials is the use of historical data because patients were not "randomized" into the two studies—the current NI trial and the "super-study" of the historical studies (see Section 12.2). For this reason, superiority testing using other designs (e.g., add-on design) is often recommended.

Although discounting may be used if the constancy assumption is violated, such a discounting is subjective. Section 5.7 makes it very clear that preservation and discounting are two different concepts, although they are indistinguishable mathematically, which is also recognized by many authors (e.g., Ng 2001, Snapinn and Jiang 2008a; Peterson et al. 2010a).

The following are highlights of the issues and challenges in the design and analysis of NI trials discussed throughout this book:

- Which metric—the difference, ratio or odds ratio—should be used in formulating the NI hypotheses with a binary endpoint (see Chapter 4)?
- Which statistical approach should be used in testing the NI hypothesis: the fixed-margin method or the synthesis method (see Section 5.5 of Chapter 5 and Section 12.4.3)?
- Should switching between NI and superiority be allowed in a confirmatory trial (see Chapter 6)?
- Should we use the confidence interval for the global mean or the predictive interval in estimating the control effect size (see Section 7.4 of Chapter 7)?
- What does the constancy assumption really mean in meta-analysis using the random-effects model (see Section 7.5 of Chapter 7)?
- What strategy should be used in the analysis of the STP design (see Section 8.2 of Chapter 8)?
- What dataset should be used in the primary analysis (see Sections 10.5 and 10.6 of Chapter 10)?
- Will indirect comparison with placebo be sufficient for approval (see Section 12.4.1)?

Other issues and challenges not covered in this book include (1) biocreep (e.g., Everson-Stewart and Emerson 2010), (2) covariate adjustment (e.g., Nie and Soon 2010) and (3) other measures of equivalence, such as the inferiority index (e.g., Li and Chi 2011). Technical details of dealing with noncompliance (see Section 10.4.3 of Chapter 10) and missing data (see Section 10.6.5 of Chapter 10) in the ITT analyses are also challenging, but they are beyond the scope of this book.

References

Dunnett CW, and Gent M (1977). Significance Testing to Establish Equivalence Between Treatments, With Special Reference to Data in the Form of 2x2 Tables. *Biometrics*, 33:593–602.

European Agency for the Evaluation of Medicinal Products, Committee for Proprietary Medicinal Products (2005). *Guideline on the Choice of the Non-Inferiority Margin* EMEA/CPMP/EWP/2158/99. http://www.ema.europa.eu/docs/en_GB/document_library/Scientific_guideline/2009/09/WC500003636.pdf (Accessed: August 25, 2013).

Everson-Stewart S, and Emerson SS (2010). Biocreep in Non-inferiority Clinical Trials *Statistics in Medicine*, 29:2769–2780.

Hassalblad V, and Kong DF (2001). Statistical Methods for Comparison to Placebo in Active-Control Trials. *Drug Information Journal*, 35:435–449.

Huitfeldt B, and Hummel J (2011). The Draft FDA Guideline on Non-Inferiority Clinical Trials: A Critical Review from European Pharmaceutical Industry Statisticians. *Pharmaceutical Statistics*. 10:414-419.

Hung HMJ, and Wang S-J (2010). Comment on PISC Expert Team White Paper: Toward a Consistent Standard of Evidence When Evaluating the Efficacy of an Experimental Treatment from a Randomized, Active-Controlled Trial. *Statistics in Biopharmaceutical Research*, 2:532–534.

Hung HMJ, Wang SJ, and O'Neil RT (2007). Issues with Statistical Risks for Testing Methods in Noninferiority Trial Without a Placebo Arm. *Journal of Biopharmaceutical Statistics*, 17, 201–213.

Hung HMJ, Wang S-J, and O'Neill R (2009). Challenges and Regulatory Experiences with Non-inferiority Trial Design Without Placebo Arm. *Biometrical Journal*, 51:324–334.

Julious SA (2011). The ABC of Non-inferiority Margin Setting from Indirect Comparisons. *Pharmaceutical Statistics*, 10:448–453.

Li G, and Chi GYH (2011). Inferiority Index and Margin in Noninferiority Trials. *Statistics in Biopharmaceutical Research*, 3:288–301.

Li T, Hutfless S, Scharfstein DO, Daniels MJ, Hogan JW, Little RJA, Roy JA, Law AH, and Dickersin K (2014). Standards Should Be Applied in the Prevention and Handling of Missing Data for Patient-Centered Outcomes Research: A Systematic Review and Expert Consensus. *Journal of Clinical Epidemiology*, 67:15-32.

Makuch R, and Simon R (1978). Sample Size Requirements for Evaluating a Conservative Therapy. *Cancer Treatment Reports*, 62:1037–1040.

Ng T-H (1993). A Specification of Treatment Difference in the Design of Clinical Trials with Active Controls, *Drug Information Journal*, 27:705-719.

Ng T-H (2001). Choice of Delta in Equivalence Testing. *Drug Information Journal*, 35:1517–1527.

Nie L, and Soon G (2010). A Covariate-Adjustment Regression ModelApproach to Noninferiority Margin Definition. *Statistics in Medicine*, 29:1107–1113.

Noninferiority Testing in Clinical Trials

Peterson P, Carroll K, Chuang-Stein C, Ho Y-Y, Jiang Q, Li G, Sanchez M, Sax R, Wang Y-C, and Snapinn S (2010a). PISC Expert Team White Paper: Toward a Consistent Standard of Evidence When Evaluating the Efficacy of an Experimental Treatment from a Randomized, Active-Controlled Trial. *Statistics in Biopharmaceutical Research*, **2**:522–531.

Peterson P, Carroll K, Chuang-Stein C, Ho Y-Y, Jiang Q, Li G, Sanchez M, Sax R, Wang Y-C, and Snapinn S (2010b). Response. *Statistics in Biopharmaceutical Research*, **2**:538–539.

Remington RD, and Schork MA (1970). *Statistics with Applications to the Biological and Health Sciences*. Englewood Cliffs, NJ: Prentice-Hall.

Röhmel J, and Pigeot I (2010). A Comparison of Multiple Testing Procedures for the Gold Standard Non-inferiority Trial. *Journal of Biopharmaceutical Statistics*, **20**:911–926.

Rothmann M, Li N, Chen G, Chi GY-H, Temple R, and Tsou H-H (2003). Design and Analysis of Non-Inferiority Mortality Trials in Oncology. *Statistics in Medicine*, **22**:239–264.

Snapinn S, and Jiang Q (2008a). Preservation of Effect and the Regulatory Approval of New Treatments on the Basis of Non-inferiority Trials. *Statistics in Medicine*, **27**:382–391.

Snapinn S, and Jiang Q (2008b). Controlling the Type 1 Error Rate in Noninferiority Trials. *Statistics in Medicine*, **27**:371–381.

Snapinn S, and Jiang Q (2011). Indirect Comparisons in the Comparative Efficacy and Non-Inferiority Settings. *Pharmaceutical Statistics*, **10**:420–426.

Snapinn S, and Jiang Q (2014). Remaining Challenges in Assessing Non-inferiority. *Therapeutic Innovation & Regulatory Science*, **48**:62–67.

Sorbello A, Komo S, and Valappil T (2010). Noninferiority Margin for Clinical Trials of Antibacterial Drugs for Nosocomial Pneumonia. *Drug Information Journal*, **44**:165–176.

U.S. Food and Drug Administration (FDA, 2004). *Transcript of Oncology Drugs Advisory Committee Meeting, July 27, 2004*. http://www.fda.gov/ohrms/dockets/ac/04/transcripts/2004-4060T1.pdf (Accessed: March 23, 2014).

U.S. Food and Drug Administration (FDA, 2010). *Draft Guidance for Industry: Non-inferiority Clinical Trials*. http://www.fda.gov/downloads/Drugs/GuidanceComplianceRegulatoryInformation/Guidances/UCM202140.pdf (Accessed: August 25, 2013).

Wang S-J, and Hung H-MJ (2003). Assessment of Treatment Efficacy in Non-inferiority Trials. *Controlled Clinical Trials*, **24**:147–155.

Index